CW00970713

THE MAHALANOBIS-TAGUCHI SYSTEM

THE MAHALANOBIS-TAGUCHI SYSTEM

GENICHI TAGUCHI

SUBIR CHOWDHURY

YUIN WU

American Supplier Institute, Inc.
United States, Brazil, Canada, Egypt, Germany, Korea, Mexico, Norway, Spain, Sweden, United Kingdom

McGraw-Hill

New York • San Francisco • Washington, D.C. • Auckland • Bogotá
Caracas • Lisbon • London • Madrid • Mexico City • Milan
Montreal • New Delhi • San Juan • Singapore
Sydney • Tokyo • Toronto

Cataloging-in-Publication Data is on file with the Library of Congress

McGraw-Hill

*A Division of The **McGraw-Hill** Companies*

1 2 3 4 5 6 7 8 9 0 DOC/DOC 0 9 8 7 6 5 4 3 2 1 0

ISBN 0-07-136263-0

The sponsoring editor for this book was Linda Ludewig and the production supervisor was Pamela A. Pelton. It was set in Fairfield by D&G Limited, LLC.

Printed and bound by R. R. Donnelley & Sons Company.

This book is printed on recycled, acid-free paper containing a minimum of 50 percent recycled de-inked fiber.

In memory of P. C. Mahalanobis

CONTENTS

PREFACE

The *Mahalanobis-Taguchi System* (MTS) is a brand-new philosophy of Genichi Taguchi. In Japan, MTS has just started revolutionizing all types of pattern recognition system such as healthcare, manufacturing, software, government etc. since its inception. These Japanese organizations have integrated this new powerful methodology in their corporate culture. The benefits these organizations have achieved are phenomenal. In the United States, the *American Supplier Institute* (ASI) already introduced this tool to Xerox Corporation, Delphi Automotive Systems, and ITT Industries, and the results in these three Fortune 500 U.S. organizations are eye-opening.

By introduction of this book, ASI wants to be a pioneer again to popularize another powerful tool in the United States. ASI is the first organization that introduced Taguchi Methods, Robust Engineering, TRIZ, and *Quality Function Deployment* (QFD) in the United States. In this groundbreaking book, we have organized 15 successfully implemented MTS case studies. One of the major important breakthroughs of MTS methodology is that it can be applied in various fields. For example, in this book successful case studies are presented from the following areas of application:

- Health Care
- Mechanical
- Electrical
- Chemical
- Space
- Software
- Government

PURPOSE OF THIS BOOK

There are no books available in any language on MTS or that present best practices with MTS. In the organizations where MTS is extremely successful, management has an understanding of the significant impact of the applications, therefore the support for implementation is highly encouraged. Our primary goal of introducing this book is:

- To contribute to the society as a whole by introducing MTS methodology. We strongly feel that MTS will be one of the most important methods of the future as pattern recognition design method.
- To provide organizations with proven techniques to become more competitive in the global market.
- To create an understanding of MTS for managers, engineers, doctors, researchers in various fields as well as academia by providing successful case studies.

IMPORTANCE OF ACTUAL CASE STUDIES

This book outlines a series of successful case studies profiling organizations that are on the cutting edge of technology and continue to lead the way with the aid of MTS. This book also provides a formula for "instant knowledge" of subject areas that apply to the reader's organization.

ABOUT THE BOOK

It is our intention to create an atmosphere of excitement for management, engineering, academia, and researchers in various fields. The first two chapters of the book outline a basic explanation and a detailed example to understand the MTS methodology via the case study technique. Chapter Three illustrates an MTS application on business process forecasting. Chapters Four through Sixteen representing seven sections, are case studies from respected companies; these studies were contributed directly by members of each organization. Each case study is unique in its application.

INTENDED AUDIENCE

This book will be a must-read for physicians, engineering managers, all types of engineers, consultants, insurance experts, financial analysts, programmers, researchers in various fields and people who have an interest in pattern recognition and forecasting. This exceptional compilation of case studies makes it ideal as a training and educational guide, as well. This book will also serve academia.

Livonia, Michigan

GENICHI TAGUCHI
SUBIR CHOWDHURY
YUIN WU

ACKNOWLEDGMENTS

The authors gratefully acknowledge the continuous support of all men and women who assisted greatly in the completion of this book:

To all of the contributors and their organizations for sharing their successful MTS case studies.

To the American Supplier Institute and its employees, especially Alan Wu and Jim Wilkins, for promoting MTS.

To Shin Taguchi for his valuable suggestions from day one.

To the Quality Engineering Society in Japan for letting us use some case studies originally published by *QES*.

To Rajesh Jugulum, a true researcher in MTS.

To Mike Hays, our publisher at McGraw-Hill, for his vision from day one.

To Linda Ludewig, our editor at McGraw-Hill, for her continuous encouragement and belief in our work.

To Esmeralda Facundo and Melissa Verduyn for their dedicated efforts during the preparation of the manuscript.

CHAPTER ONE

INTRODUCTION

1.1 LIVES VERSUS MACHINES

Many people remember the logo of the RCA Victor Company: a dog listening to his master's voice over a phonograph. Obviously, a dog recognizes his master's voice. Similarly, painting lovers can easily recognize the impressionist paintings of Vincent van Gogh. Classical music lovers can also distinguish the music composed by Johann Sebastian Bach from other composers' pieces. But even when people use the most advanced computers available today, it is not easy for man-made machines to recognize the human voice, faces, handwriting, or fingerprints.

One of the reasons why we have the capability to recognize a pattern is because about nine million optical-sensing cells exist in our eyes. Assuming that the time required to treat the captured image is four-hundredths of a second, we can treat 225,000,000 pieces of information in one second. Another reason is because we do not memorize a vast amount of information; rather, we efficiently simplify or screen out unnecessary parts of information. If we could develop a system that is simple enough to use current personal computers for pattern recognition, there would be numerous applications in various areas.

1.2 WHAT IS MTS?

MTS stands for the Mahalanobis-Taguchi System. Professor P. C. Mahalanobis was a famous Indian statistician who established the *Indian Statistical Institute* (ISI), and Genichi Taguchi is famous for the concept of robust engineering. In 1930, Mahalanobis introduced a statistical tool called Mahalanobis Distance, which is used to distinguish the pattern of a certain group from other groups—much like the process by which a doctor determines whether or not a patient has a certain kind of disease.

Mahalanobis applied his Mahalanobis Distance technique to archaeology in order to distinguish whether excavated bones were the bones of an elephant by using his statistical test of significance. In other words, Mahalanobis Distance is a process of distinguishing one group from another. Taguchi introduced Mahalanobis Distance into the process of defining a reference group and measuring individual subsets. For example, we could use the Mahalanobis Distance technique to measure a person's degree of health at the time of a health examination ranging from healthy to severely ill person by using all available multidimensional data.

For the healthy group (subset) of the population, we can assume that the Mahalanobis Distance is one scalar number calculated from the data and averages the pattern distance of the healthy group. Outside of the group, we expect that the pattern gradually changes, creating a larger distance from the zero point.

If the distance measurement agrees with a doctor's judgment about the severety of a person's illness, then we may use the distance as a measure of the level of the individual's health. The measure of accuracy of Mahalanobis Distance against doctors' judgments is the *Signal-to-Noise* (SN) ratio, which is the most important scale measurement of the Taguchi Methods.

Taguchi also provides the systematic steps for parameter design, the selection of the reference group, and the selection of the data's variables using an orthogonal array and the SN ratio. Although one pattern-recognition method is Mahalanobis Distance, the SN ratio is the measure of accuracy of the Mahalanobis Distance measurement, which is used to evaluate the method.

MTS incorporates the three strategic methods in information system design. The first strategy is to introduce only one measurement scale in any multidimensional space by using the Mahalanobis Distance in any subset of the selected space as *uniform one* and calculating the distance from the norm against the distance of other members. The second strategy is to use the SN ratio of the distance against the number of the space that we know as the true value or the true classification. The third strategy is to optimize all factors of the information to improve the SN ratio with an orthogonal array.

MTS is a new approach developed by Taguchi that applies Mahalanobis Distance to various areas. Basically, MTS is a measuring or evaluating tool used to recognize a pattern from multidimensional data. In MTS, the quality of measurement is evaluated by the SN ratio.

1.3 WHERE CAN MTS BE APPLIED?

MTS can be applied to two major objectives: diagnosis and forecasting. The following paragraphs describe the areas of application.

PATIENT MONITORING

In a conventional medical checkup, diagnoses are made by the experience of a doctor along with the results of testing. Usually, many test items exist—and each item has a range. A person will be judged as to whether the tested result of one item falls beyond a certain range. Then, he or she will be sent for a close examination by increasing the number of check items. In biochemical testing, normal values (in general) are arbitrarily determined by the test-chemical manufacturers, and in extreme cases, textbook values are used without modification.[1]

Many of those test items are correlated. In the application of MTS, the test results from a healthy group are collected in

[1]Kanetaka, Tatsuji. "Diagnosis of a Special Health Check Using Mahalanobis Distance," *ASI Journal* Vol. 3, No. 1., 1990.

order to construct a Mahalanobis Space for diagnosis. This approach can be used for all kinds of diagnoses and can also be used to predict the amount of time necessary for a patient to recover.

Perhaps the most exciting potential application of MTS is researching medicines or medical treatment. Currently, a newly developed medicine is evaluated by so-called double-blind tests that use two groups. One group is given the test drug, and another group is given a placebo. This test requires a large number of people and is inhuman for the group that is taking the placebo. The change in Mahalanobis Distance can forecast the effect of the new drug or the new treatment method and can possibly use only one person in a short period of time.

MANUFACTURING

Pattern recognition is widely used in manufacturing. For example, many inspection works are done by eyeball observation, such the appearance of welded or soldered component parts. When the fraction of defective products in a production line is low and the worker has to inspect 300 pieces per hour, it is easy to overlook a defective product. Many successful studies from mechanical, electrical, and chemical industries have been reported.

FIRE DETECTION

The law requires fire-alarm systems to be installed in public buildings and hotels. Perhaps most of us have experienced a false alarm in which the alarm sounds, but there is no fire. A smoke detector might sound because of cigarette smoking or barbecuing.

In the development of fire-alarm systems, data such as temperature or the amount of smoke are collected from the situations without a fire. These data are considered a reference group for the construction of the Mahalanobis Space.

EARTHQUAKE FORECASTING[2]

In the case of earthquakes, knowing what kind of data should be collected is difficult. The method of data collection depends on what type of forecasting will be made. For example, we might try to forecast an earthquake one hour from now, but such forecasting might not be too useful (except for the fact that we could take some preventive actions, such as shutting off the gas valve).

Suppose that we are going to forecast an earthquake that will occur 24 to 48 hours from now. We must know that in MTS, the data to be collected are under normal conditions. In medical checkups, the data are collected from the healthy people. In fire detection, data are collected under the "no fire" condition. Likewise, the data "without earthquake" must be collected as a reference group. Weather bureaus have the records from seismographs for different years. The data "48 hours without an earthquake" is collected in order to construct a Mahalanobis Space. Then, the data "48 hours right after an earthquake" are collected and compared for the study of future forecasting.

WEATHER FORECASTING[3, 4]

MTS could be used for weather forecasting—not to discuss current forecasting methodology itself, but to provide an approach that summarizes all kinds of existing data and simplifies the process by reducing the amount of information that is not relevant.

[2]Taguchi, Genichi. "Earthquake Prediction and Quality Engineering," Quality Engineering Forum, Vol. 3, No. 3, 1995.

[3]Taguchi, Genichi. "Weather Forecasting System Design with Mahalanobis Distance," Quality Engineering Forum, Vol. 3, No. 6.

[4]Taguchi, Genichi. "Weather Forecasting and Quality Engineering," Quality Engineering Forum, Vol. 4, No. 1.

AUTOMOTIVE COLLISION PREVENTION SYSTEM

Automotive manufacturers are developing sensors that can detect a dangerous condition during driving. In the case of airbags, we want to avoid both types of errors: the airbag being activated when there was no crash, and the airbag not activating at the moment of a crash. In order to construct a Mahalanobis Space, the acceleration at several locations in a car are recorded from time to time, probably at an interval of a few thousandths of a second. Such data are collected during various driving conditions, such as driving on an extremely rough road, quick acceleration, or a quick stop. Those data are used with other sources of information, such as driving speed, in order to construct a Mahalanobis Space. The computer that controls the airbag system is constantly calculating a Mahalanobis Distance. When the distance becomes greater than a predicted threshold, the system sends a signal to activate the airbag(s) before the crash. An early stage of research has been reported.[5]

BUSINESS APPLICATIONS

Prediction is one of the important areas of pattern recognition. In the business arena, banks or credit-card companies make predictions for loan or credit-card approval based on the application form or the information obtained from credit companies. A Mahalanobis Space can be formed from the existing good customers, and Mahalanobis Distance can be used to predict creditworthy applicants.

OTHER APPLICATIONS

MTS can be applied in many other areas, such as fingerprinting, handwriting analysis, and character or voice recognition. Although there are some products of this kind already on the

[5]Mizoguchi, Kazutaka. "A Research on a Sensing System Using Mahalanobis Distance for Prevention of Driving Accidents," Quality Engineering Forum Symposium, 1998.

market, they are far from mistake free. The use of Mahalanobis Distance could contribute to further improvement.

1.4 AN IMPORTANT ISSUE

MTS basically involves one scale type of measurement. In measurement, there are two requirements: the origin and the scale. In MTS, the origin is zero. The scale to be used is Mahalanobis Distance.

The ideal function in measurement is that measured value is proportional to true value; therefore, a dynamic SN ratio is used. The reciprocal of the SN ratio is the error variance after the calibration of a measuring system. When the dynamic SN ratio is applied to MTS, the reciprocal of the SN ratio is the error variance of misjudgment. In MTS, you should note that the dynamic SN ratio must be used, rather than discussing two types of errors or chi-square distribution. You should utilize Mahalanobis Distance in continuous mode, rather than in discrete mode.

In medical treatment, for example, the Mahalanobis Distance of a certain patient was calculated from medical test results before the patient was treated. After a certain period of treatment, another Mahalanobis Distance was calculated from the new test results. If it is known that the larger the Mahalanobis Distance, the more severe the illness, then the value that is calculated from the new test should become smaller if the illness improved. A case study in this regard has been published in order to show the usefulness of this approach. MTS has an extremely important forecasting use in this area. From the relationship between Mahalanobis Distance and the time of treatment, a doctor can predict when the patient can leave the hospital.

That point indicates another important issue. Medical research, such as researching the effect of a new treatment or a new drug, can be made by using just one patient, instead of relying on so-called double-blind tests where many people are studied—with half being untreated or treated by a placebo.

Sometimes the study takes years. By using the dynamic SN ratio, researchers can observe the trend more quickly in a short period of time. Such applications have a huge potential for future research and would bring immeasurable benefits to society.

Many case studies introduced in this book, however, use a "larger-the-better" type of SN ratio for two reasons. One reason is because of convenience, because it is easier to understand and easier to calculate. Another reason is because the true values that are needed to calculate the dynamic SN ratio are unknown in many cases. This area must be further developed in the future. Nevertheless, the case studies in this book show much success. They are useful for communicating that MTS can be applied in various areas.

CHAPTER TWO

A DETAILED EXAMPLE OF MTS

2.1 CLUTCH DISC DEFECTS

Clutch discs used in automobiles are designed in different shapes and dimensions for various torque transformation requirements. Figure 2-1 shows one example.[1] The disc with an 180 mm outside diameter and 15 mm width is glued with a special frictional material on both sides, and a slit-type pattern is engraved on the surface. The quality of discs is strictly controlled, because the quality affects driving performance.

FIGURE 2-1 A clutch disc

[1]Teshima, Shoichi, Bando Tomonoyi, Jin Dan (Japan Technical Software Co.). "A Research of Defects Detection Using Mahalanobis Taguchi System," Journal of Quality Engineering Forum, Vol. 5, No. 5.

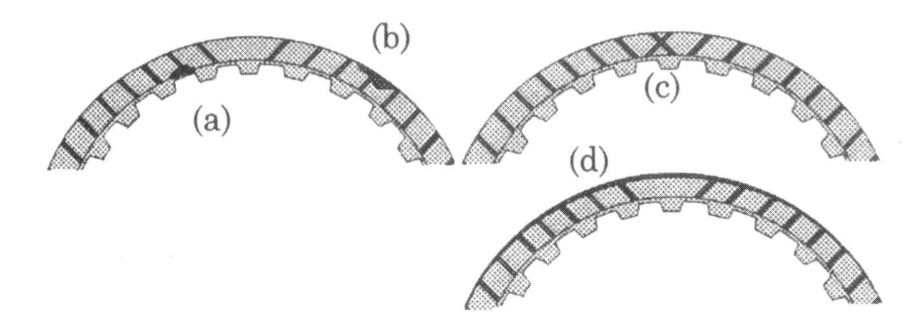

FIGURE 2-2 Appearance of clutch discs

Figure 2-2 shows four defect types to be inspected:

(a) Glue adhesion—The glue sticking to the surface or to the edge
(b) Frictional material peel-off
(c) Pattern misalignment
(d) Friction material eccentricity

2.2 VISUAL INSPECTION METHOD

In order to automate the inspection, you must design a system that can summarize multi-dimensional information. Such a system does not have to use the same quality characteristics as human inspection, as long as the automatic inspection gives the same results.

The patterns are obtained while rotating a clutch disc at a constant speed, and a line *Charge Coupled Device* (CCD) camera takes pictures, as shown in Figure 2-3. Such patterns are inputted into a personal computer that has specially designed image-input boards. When a disc rotates once, lengthy image data (disc width × one rotation) can be obtained. To take the picture, you can randomly select any starting point on the disc.

The image data represent the light concentration of images. Figure 2-4 shows the wave pattern of one rotation. The abscissa

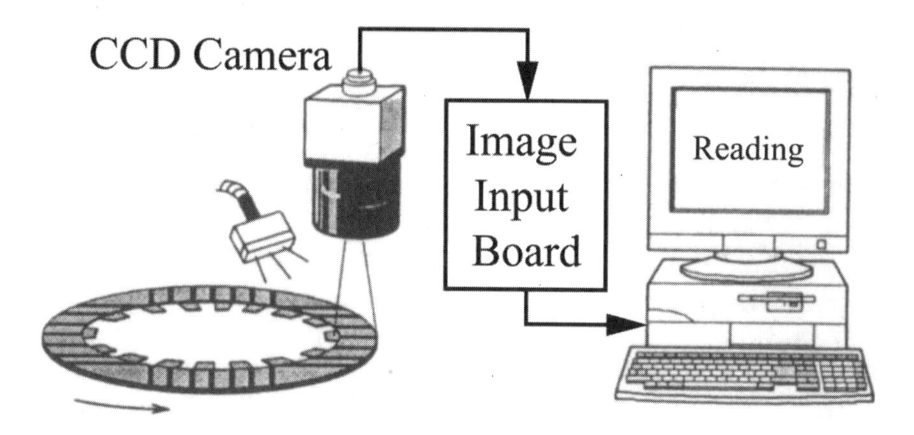

FIGURE 2-3 Pattern-collection device

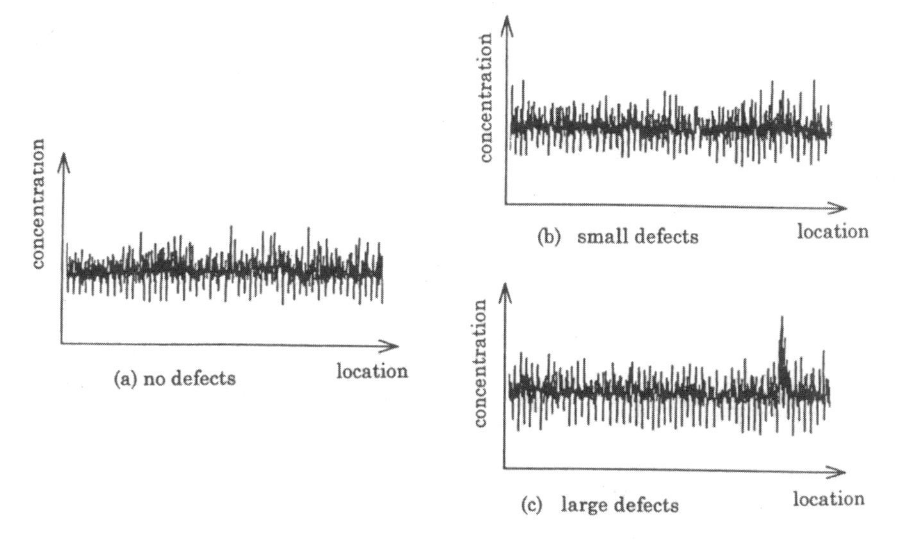

FIGURE 2-4 Examples of patterns

represents the rotating position, and the ordinate is the light concentration.

(a) is the pattern from a defect-free product. (b) represents glue adhesion, and (c) shows the friction material peeled off. Although the deviation of pattern (c) from the normal pattern (a) is explicit, the human eye cannot recognize pattern (b).

Therefore, recognizing both the implicit deviation and the explicit deviation is the main goal.

2.3 DIFFERENTIAL AND INTEGRAL CHARACTERISTICS

For pattern recognition, Taguchi suggests using differential and integral characteristics. In Figure 2-5, Y(t) denotes one wave (the wave obtained from one-disc rotation), and *p* parallel lines on the t-axis with the same intervals toward the Y-axis direction are drawn. Next, the number of cross-sections that the wave

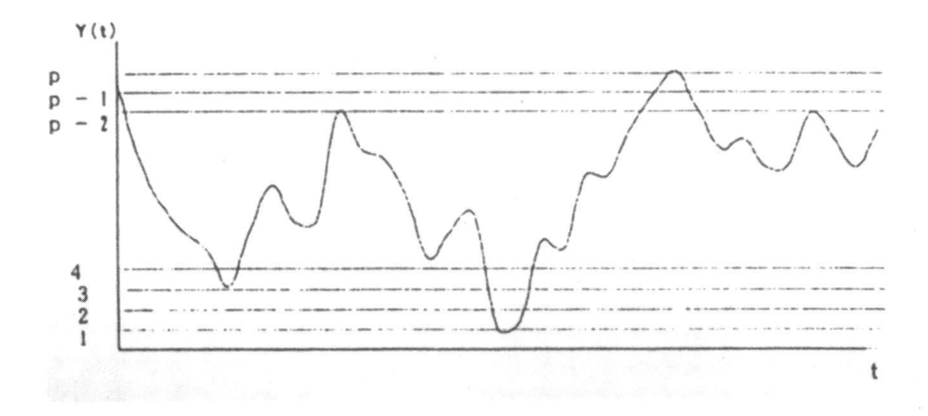

P	Differential	Integral	A	B	C	D
1	18	5989	75	1318	54	66
2	46	5971	68	1098	82	66
3	72	5944	65	828	92	65
.						
.						
37	8	4	1	1	407	1593
38	8	4	1	1	407	1593
39	4	2	1	1	1525	2552
40	2	1	1	1	2552	3447

Note: Row numbers show parallel line numbers

FIGURE 2-5 Differential and integral characteristics from the wave pattern

intersects on each line is observed as the differential characteristic. Then, the range between cross-sections above each line represents the integral characteristic.

The differential characteristic indicates wave changes, similar to the concept of frequency. The differential characteristic is observed for each amplitude, so that the frequency distribution on the amplitude direction is measured. The integral characteristic indicates the magnitude of the range of each amplitude. These two characteristics are much easier to collect and manipulate than traditional frequency characteristics. Because these characteristics include information regarding frequency, amplitude, and distribution, they can be used for quick and efficient data processing.

Besides these two characteristics, four non-disclosed characteristics were used in this study.

2.4 RAW DATA COLLECTION

From normal clutch discs, 1,000 sets of data were generated in order to show images. These clutches were verified as defect free. A set of data from one disc is a collection of 36 waves taken in the radial direction.

Each wave contains a total of 240 characteristics, including differential and integral characteristics and four non-disclosed characteristics for 40 parallel lines. From 240 characteristics, 160 characteristics were selected based on the conclusion of a preliminary experiment. Therefore, there were 1,000 sets of 160 characteristics; thus, from each wave, a correlation matrix and its inverse matrix were calculated.

Because there were 36 waves in each clutch, 36 sets of correlation matrices and their inverse matrices were calculated as follows:

One disc—36 waves

One wave—40 parallel lines

Each line—six characteristics (differential, integral, and four non-disclosed characteristics)

Total number of characteristics in one wave—$40 \times 6 = 240$ (reduced to 160)

Table 2-1 shows examples of characteristics. In this table, row numbers represent parallel line numbers.

As described previously, 160 characteristics were selected from 240 characteristics in Table 2-1. Let those characteristics be denoted by the following:

$Y_1, Y_2, \ldots Y_{160}$

Table 2-2 shows part of the data from one wave.

2.5 NORMALIZATION

In Table 2-2, there are 160 characteristics with 1,000 good products. Because the 160 characteristics have different scales, the raw data are standardized by using Equation (2-1) to obtain Table 2-3.

TABLE 2-1 Examples of Characteristics

P	DIFFERENTIAL	INTEGRAL	A	B	C	D
1	18	5989	75	1318	54	66
2	46	5971	68	1098	82	66
3	72	5944	65	828	92	65
.	.	.	.	.	.	.
.	.	.	.	.	.	.
.	.	.	.	.	.	.
37	8	4	1	1	407	1593
38	8	4	1	1	407	1593
39	4	2	1	1	1525	2552
40	2	1	1	1	2552	3447

Note: Row numbers show parallel line numbers.

TABLE 2-2 Raw Data

Characteristics	Y_1	Y_2		Y_i		Y_j		Y_{160}
1	$Y_{1.1}$	$Y_{2.1}$	......	$Y_{i.1}$	......	$Y_{j.1}$	......	$Y_{160.1}$
2	$Y_{1.2}$	$Y_{2.2}$	......	$Y_{i.2}$	......	$Y_{j.2}$	......	$Y_{160.2}$
⋮	⋮	⋮	⋮	⋮	⋮	⋮	⋮	⋮
l	$Y_{1.l}$	$Y_{2.l}$	......	$Y_{i.l}$	......	$Y_{j.l}$	......	$Y_{160.l}$
⋮	⋮	⋮	⋮	⋮	⋮	⋮	⋮	⋮
1000	$Y_{1.1000}$	$Y_{2.1000}$	......	$Y_{i.1000}$	......	$Y_{j.1000}$	......	$Y_{160.1000}$
Total	Y_1	Y_2	......	Y_i	......	Y_j	......	Y_{160}
Average	$\overline{Y}_1$	$\overline{Y}_2$	......	$\overline{Y}_i$	......	$\overline{Y}_j$	......	$\overline{Y}_{160}$
Standard Deviation	σ_1	σ_2	......	σ_i	......	σ_j	......	σ_{160}

TABLE 2-3 Standardized Data

Characteristics	Y_1	Y_2		Y_i		Y_j		Y_{160}
1	$y_{1.1}$	$y_{2.1}$	......	$y_{i.1}$	......	$y_{j.1}$	......	$y_{160.1}$
2	$y_{1.2}$	$y_{2.2}$	......	$y_{i.2}$	......	$y_{j.2}$	......	$y_{160.2}$
⋮	⋮	⋮	⋮	⋮	⋮	⋮	⋮	⋮
l	$y_{1.l}$	$y_{2.l}$	......	$y_{i.l}$	......	$y_{j.l}$	......	$y_{160.l}$
⋮	⋮	⋮	⋮	⋮	⋮	⋮	⋮	⋮
1000	$y_{1.1000}$	$y_{2.1000}$	......	$y_{i.1000}$	......	$y_{j.1000}$	......	$y_{160.1000}$

$$y_{il} = \frac{Y_{il} - \overline{Y}_i}{\sigma i} \tag{2.1}$$

2.6 MATRIX CONSTRUCTION

From Table 2-3, the correlation coefficient of each pair of characteristics, denoted by r_{ij}, is calculated by using Equation (2-2), where n is the number of observations of each characteristic. In this case, n is equal to 1,000.

$$r_{ij} = \frac{(\Sigma y_{il} \times y_{jl})}{n} \qquad (l=1, 2, \cdots, n) \tag{2.2}$$

From these results, their correlation coefficient matrix is constructed as shown in Equation (2.3).

$$R = \begin{bmatrix} 1 & r_{12} & \cdots & r_{1k} \\ r_{21} & 1 & \cdots & r_{2k} \\ \cdot & \cdot & \cdot & \cdot \\ \cdot & \cdot & \cdot & \cdot \\ \cdot & \cdot & \cdot & \cdot \\ \cdot & \cdot & \cdot & \cdot \\ \cdot & \cdot & \cdot & \cdot \\ \cdot & \cdot & \cdot & \cdot \\ r_{k1} & r_{k2} & \cdots & 1 \end{bmatrix} \tag{2.3}$$

where k is the number of characteristics. In this case, k is equal to 160. We can see from Equation (2.3) that the values in the diagonal line are equal to 1.

From the correlation matrix, its inverse matrix, denoted by A (or R^{-1}), is constructed as shown in Equation (2.4).

$$A = R^{-1} = \begin{bmatrix} a_{11} & a_{12} & \cdots & a_{1k} \\ a_{21} & & & \\ \cdot & \cdot & \cdot & \cdot \\ \cdot & \cdot & \cdot & \cdot \\ \cdot & \cdot & \cdot & \cdot \\ \cdot & \cdot & \cdot & \cdot \\ \cdot & \cdot & \cdot & \cdot \\ \cdot & \cdot & \cdot & \cdot \\ a_{k1} & a_{k2} & \cdots & a_{kk} \end{bmatrix} \tag{2.4}$$

2.7 MAHALANOBIS DISTANCE

From Table 2-3, a Mahalanobis Distance denoted by D^2 is calculated by using Equation (2.5).

$$D^2 = \frac{\Sigma a_{ij}\, y_i\, y_j}{k} \qquad (2.5)$$

These values construct the Mahalanobis Space. According to Mahalanobis' Law, if all of those products are defect free (or belong to the normal group or to the reference group), the average of these values approximately follow chi-square distribution with a degree of freedom of *k*.

The core of statistics is distribution. In quality engineering, however, distribution is not considered. Instead, the SN ratio is calculated to evaluate the magnitude of diagnosis error, which can be calculated without considering distribution.

The average of D^2s in Equation (2.5) is equal to 1 if the raw data were collected from the reference group or from the defect-free group. If the pattern of a disc differs from the reference pattern, the Mahalanobis Distance of this disc becomes extremely large.

2.8 RESULTS OF RECOGNITION

Figure 2-6 shows examples of recognized results: (a), (b), (c), and (d). Each figure represents the results from one disc containing 36 Mahalanobis Distances calculated from 36 waves in one disc. The abscissa shows different positions toward radius direction, and the coordinate shows Mahalanobis Distances.

(a) shows the results from the basic data, and their distances are extremely small. (b) shows the data taken from a nondefected disc, and their distances are less than 2. (c) shows a disc that has a small amount of glue on the surface. Many positions have values larger than 4, suggesting that the disc is not normal. (d) shows a disc where the abrasion material was peeled off, obviously showing that the disc is defective. In addition to these examples, various discs were inspected. We confirmed that the results agreed with human inspection.

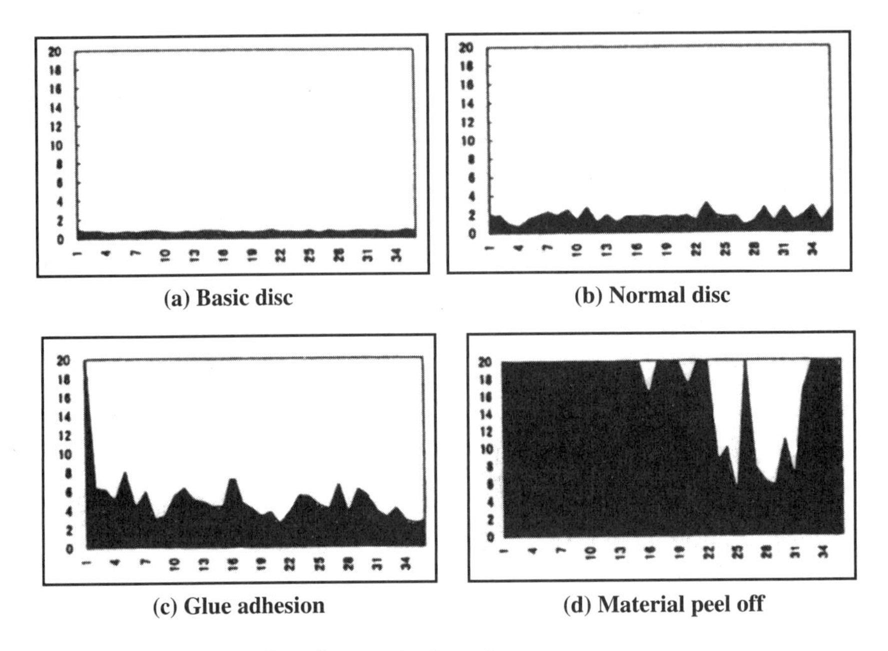

FIGURE 2-6 Examples of recognized results

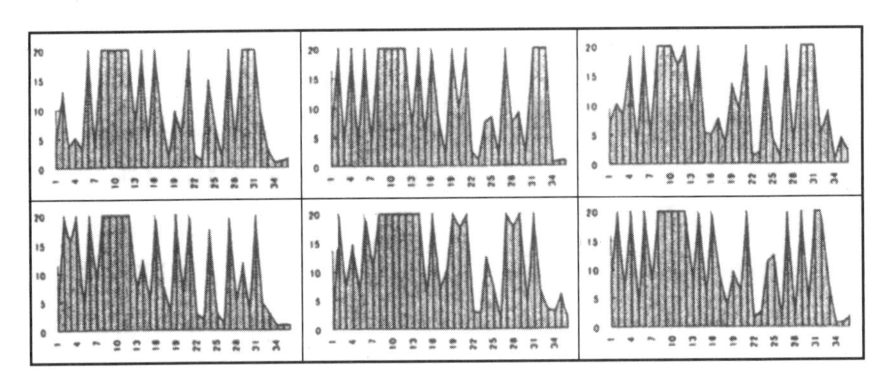

FIGURE 2-7 Comparison of different starting positions of image taking

Figure 2-7 shows the results of a defective disc. The images were taken from six different starting positions on the disc. All six figures are similar, indicating that the results of recognition are not affected by different starting positions.

2.9 SELECTION OF A THRESHOLD VALUE

Because the distance calculated from a reference group is small and the distance from the abnormal group is large, you must select a threshold value to judge whether the distance from a sample belongs to the reference group. When the selected threshold is too large, the mistake of judging a bad threshold value as a good value becomes large, and vice-versa.

In quality engineering, the right way to determine the threshold value is to find the point at which the losses due to the two types of mistakes are balanced. In this case, 4 or 5 can be used as the threshold.

2.10 SELECTION OF CHARACTERISTICS

We had 160 characteristics in the experiment. In order to reduce the cost and speed up data processing, we had to screen out the characteristics that were not contributing to the experiment. An L_{64} orthogonal array was used for this purpose.

Forty out of 160 characteristics were selected as important, and these characteristics were used all of the time and were not considered for screening. The other 120 characteristics to be screened were separated into three groups. Each group was to be studied separately, and three experiments were conducted to study 40 characteristics each.

EXPERIMENT 1

The 40 characteristics in the first group were assigned to an L_{64} orthogonal array. For each characteristic in a given column, Level 1 represented "not using" this characteristic, and Level 2 represented "using" this characteristic. The other 80 characteristics were used with the 40 characteristics that were to be used all of the time. From each run of the orthogonal array, a correlation matrix and its inverse matrix were calculated from the base data.

Ten discs having the glue adhesion and abrasion material peel-off were used to determine Mahalanobis Distances. Because these distances were "larger-the-better," the following SN ratio, η, was used for analysis:

$$\eta = -10\log \frac{1}{10} \left(\frac{1}{D_1^2} + \frac{1}{D_2^2} + \dots + \frac{1}{D_{10}^2} \right) \tag{2.6}$$

where *m* is the number of data points. In this case, *m* is equal to 10.

Table 2-4 shows the L_{64} orthogonal array and the calculated SN ratios. The factorial effect of each characteristic was calculated as shown in Figure 2-8.

Factorial effect = (total of SN ratios of Level 2)
 − (total of SN ratios of Level 1)

From these results, the larger 20 characteristics were selected; thus screening out the 20 less-relevant characteristics.

TABLE 2-4 The orthogonal array L_{64} and SN ratios

	1	2	3	4	5	6	7	8	9	10	....	56	57	58	59	60	61	62	63	SN η
1	1	1	1	1	1	1	1	1	1	1		1	1	1	1	1	1	1	1	1.54
2	1	1	1	1	1	1	1	1	1	1		2	2	2	2	2	2	2	2	1.63
3	1	1	1	1	1	1	1	1	1	1		2	2	2	2	2	2	2	2	1.98
4	1	1	1	1	1	1	1	1	1	1	....	1	1	1	1	1	1	1	1	2.08
5	1	1	1	1	1	1	1	2	2	2		2	2	2	2	2	2	2	2	2.2
6	1	1	1	1	1	1	1	2	2	2		1	1	1	1	1	1	1	1	2.3
⋮						⋮					⋱				⋮					
54	2	2	1	1	2	2	1	2	1	1		1	2	2	1	1	2	2	1	2.11
55	2	2	1	1	2	2	1	2	1	1		1	2	2	1	1	2	2	1	2.12
56	2	2	1	1	2	2	1	2	1	1		2	1	1	2	2	1	1	2	2.08
57	2	2	1	2	1	1	2	1	2	2		1	2	2	1	2	1	1	2	2.21
58	2	2	1	2	1	1	2	1	2	2	....	2	1	1	2	1	2	2	1	2.15
59	2	2	1	2	1	1	2	1	2	2		2	1	1	2	1	2	2	1	2.22
60	2	2	1	2	1	1	2	1	2	2		1	2	2	1	2	1	1	2	2.16
61	2	2	1	2	1	1	2	2	1	1		2	1	1	2	1	2	2	1	2.15
62	2	2	1	2	1	1	2	2	1	1		1	2	2	1	2	1	1	2	2.11
63	2	2	1	2	1	1	2	2	1	1		1	2	2	1	2	1	1	2	2.17
64	2	2	1	2	1	1	2	2	1	1		2	1	1	2	1	2	2	1	2.14

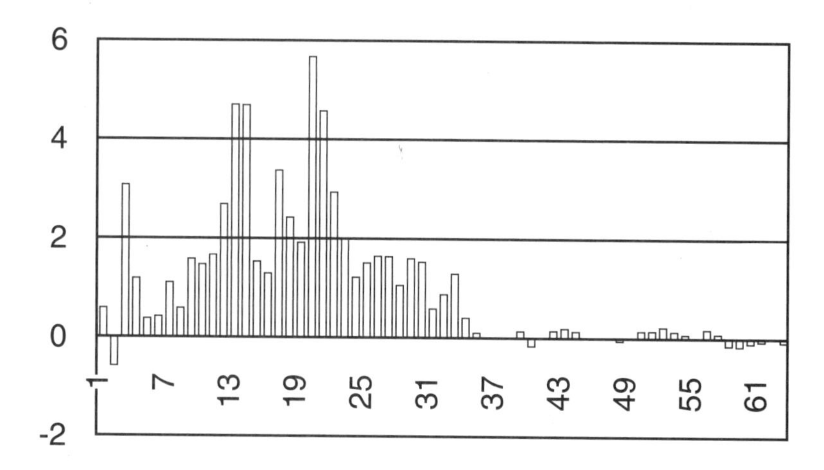

FIGURE 2-8 Factorial effects

EXPERIMENTS 2 AND 3

Similar experiments were conducted for the characteristics in group 2 and group 3. The larger 20 characteristics from each group were selected.

2.11 CONFIRMATION

The number of characteristics were reduced from 160 to 100: the total of 40 (pre-selected), 20 (Experiment 1), 20 (Experiment 2) and 20 (Experiment 3). To confirm the results, these 100 characteristics were used in the following experiment. Figure 2-9 shows the comparison of SN ratios between the cases using 160 and 100 characteristics, respectively, inspecting defective discs with glue adhesion. The higher and lower lines represent these two cases.

The higher line, which represents using 100 characteristics, shows a better discrimination—possibly due to the reduction of noise by eliminating non-contributing characteristics.

We confirmed that the inspection to discriminate normal and abnormal products can be effectively made by using the

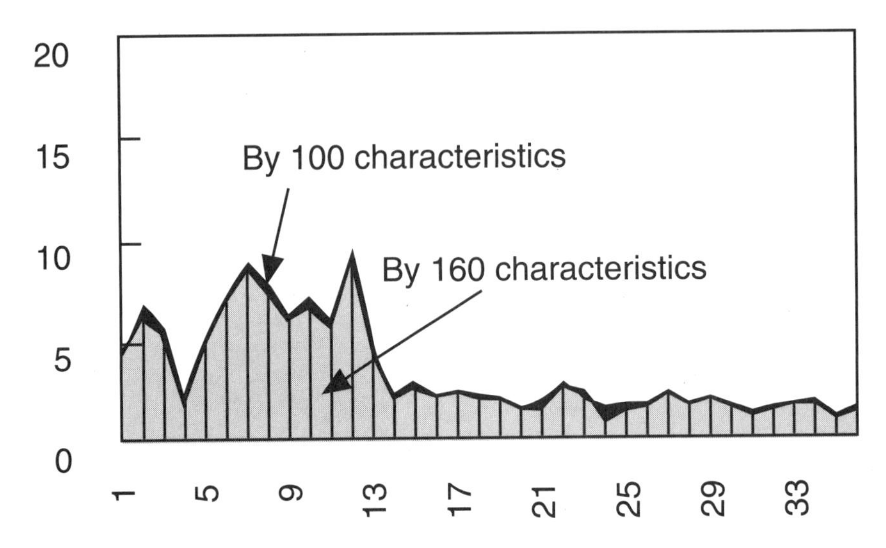

FIGURE 2-9 Comparison before and after screening

MTS method through wave patterns (by using differential and integral characteristics). Also, the number of characteristics could be reduced to only those that significantly contribute to recognition by using orthogonal array experiments.

As a result, the time required for inspection could be reduced to three seconds, which is practical for the ongoing inspection process. We also confirmed that the brightness variation for inspection does not affect results. No human errors would exist, and furthermore, this method could detect problems such as abrasive material eccentricity, which is difficult to detect with the human eye.

CHAPTER THREE

BUSINESS-PROCESS FORECASTING

3.1 A PROBLEM IN BUSINESS SYSTEMS

The first industrial revolution liberated the human from heavy labor through the mechanization of agricultural processing or transporting. This revolution also raised the living standard by improving productivity. In the United States, for example, the farmers, who represented fewer than 2 percent of the entire population, could produce twice the amount of agricultural product that was needed in the country. Such an improvement was achieved by the mechanization of the farmer's work, transportation, and storage.

In today's society, much work must still be done by humans. Taking care of babies or the elderly are typical examples. We want to enable machines to work like humans—to design the robots that have intelligence like humans but that have no selfish desires like some humans have. One example of the human (or animal) intelligence that is lacking in computers is the capability of recognizing patterns. In a broad sense, pattern recognition is the capability for language comprehension. In this chapter, we discuss the issue of making judgments from a database by using a computer.

Taguchi methods include a series of methods that use the *Signal-to-Noise* (SN) ratio to evaluate the functionality of products or systems. In pattern recognition, the SN ratio is used to evaluate the magnitude of the error in judgment in the same way that it is used to evaluate the magnitude of the measurement error. The SN ratio is applicable not only to where functionality is discussed, but also to multi-dimensional space. In this chapter, we describe the SN ratio for evaluation of the ability to judge or forecast in multi-dimensional space.

In multi-dimensional space, various types of time-series data can also be included; therefore, this approach is applicable to many different fields. Data must be collected by the specialists in each field, however, and not by those who are specialized in Taguchi methods. Taguchi methods are a series of universal technologies or common technologies, as opposed to the technology in a specialized field, such as electrical engineering or finance. Taguchi methods provide the following two approaches to treat multi-dimensional databases:

- Evaluation of error in a pattern-recognition process by using the SN ratio, based on Mahalanobis Distance as the base space
- A method to improve the efficiency of a pattern-recognition process through the simplification of data

3.2 PREPARATION OF THE DATABASE

In a measuring system, there are two essential elements: a base point and a scale. In order to define the base point and the scale of a multi-dimensional space in order to construct a Mahalanobis Space, the data used in the base space are important.

In the case of credit cards, the data to be used as a base are the group of customers who pay bills. Such a group is called the normal group. The group includes those who did not pay on time but will pay later when they are prompted. These people have the ability to pay but did not pay on time, because cash was not available when the payment was due. Such people are really creditable people, although the balance in their savings was not enough.

In the United States, most credit limits vary from about $5,000 to $30,000. A customer who has such a line of credit can spend the money without having the equivalent amount of savings in the bank. Many financing organizations in Japan approve loans only to customers who have savings accounts. Another option is asking for collateral. Such institutions have no capability to generate income from loans and have no capability to judge clients. Having the capacity to find the customers who are able to pay without collateral is important. Using traditional concepts to determine the amount of a loan is not always successful. The following list describes a method of evaluating credit by using the SN ratio.

In order to construct a Mahalanobis Space for loans, various types of data are collected from a person or from a company:

- sex
- type of work
- past three years' income
- past three years' taxes paid
- residence status
- family status
- education
- past three years' expenses (if available)
- other factors

No limit exists on the number of items to be collected. From all cardholders, select only the people who did pay as the normal group. Defining who belongs to the normal group and how to construct the database are the two most important tasks. The reason why traditional studies were not successful in judgment is because they only studied the people who did not pay. Instead, the data to be collected must be from the normal group—the group who pays. This group should be used as a base, much like the case of health diagnoses—where data are collected from the people who are healthy, instead of from the people who are sick.

Once data are collected, a statistic is constructed. First, data are classified into two categories: continuous variables and classified attributes. Classified attributes are treated in one of the following ways:

- Convert into 0 or 1 data
- Give points to each class if there is an order

For example, if the variable "gender" is denoted by x_1, then "male" and "female" are denoted by 0 and 1, respectively.

male: $x_{11} = 0$
female: $x_{12} = 1$

If the variable "occupation" is denoted by x_2, the conversion is made as follows:

self-employed: $x_{21} = 0$
employee: $x_{22} = 1$

When there are more classes, occupation is treated as a series of multi-variables denoted by x_i. Those data can be converted as follows:

public official?	yes: $x_2 = 0$	no: $x_2 = 1$
listed company employee?	yes: $x_3 = 0$	no: $x_3 = 1$
non-listed company employee?	yes: $x_4 = 0$	no: $x_4 = 1$
medical doctor?	yes: $x_5 = 0$	no: $x_5 = 1$
lawyer?	yes: $x_6 = 0$	no: $x_6 = 1$
small-scale self employer?	yes: $x_7 = 0$	no: $x_7 = 1$
large-scale employer?	yes: $x_8 = 0$	no: $x_8 = 1$
housewife?	yes: $x_9 = 0$	no: $x_9 = 1$
others?	yes: –	no: –

Continuous variables are directly used in the calculation without any conversion. Once all data are collected, the Mahalanobis Space is fixed. You should collect enough data for at least 1,000 people who paid their dues as the reference group, regardless of whether the nature of the data consists of 0 and 1 or continuous variables. Again, having a correct database is the most important, because it determines the ability of judgment.

A credit-card company can freely create its own database that shows its capability of judgment. Those items or variables included in the database are called the control factors for pattern recognition, such as eight variables to describe the occupation variable in the previous example. Instead of considering eight variables, the company might also give points to different occupations. If those points are given inappropriately, the SN ratio calculated from these results will show a poor judging capability. Besides the surveyed data, someone in the company might also give points to individuals as another variable. Again, the type of control factors that a company should use is entirely at its own discretion. Thanks to modern, powerful computers, it is possible to use up to 1,000 control factors in order to construct a Mahalanobis Space.

3.3 CALCULATION OF THE DATABASE

Let the number of people in the normal group who paid be represented by *n*, and let the number of variables in the database be *k*. The surveyed results are shown in Table 3-1.

From each variable, the mean m_i and the standard deviation σ_i are calculated.

$$m_i = \frac{1}{n}(X_{i1} + X_{i2} + \dots + X_{in}) \tag{3.1}$$

$$(i = 1,2,\dots,k)$$

TABLE 3-1 Surveyed Results

Person	1,	2,	....,	n
Variable	X_{11}, X_{21}, . . . X_{k1},	X_{12}, X_{22}, X_{k2},	...., ..., ...,	X_{1n} X_{2n} X_{kn}

$$\sigma_i = \sqrt{\frac{1}{n-1}[(X_{i1} - m_1)^2 + \dots + (X_{in} - m_i)^2]} \qquad (3.2)$$

$(i = 1, 2, \dots, k)$

From the results, the following normalized variables, X_{il} $(i = 1, 2, \dots, k, l = 1, 2, \dots, n)$ are calculated.

$$x_{il} = \frac{X_{il} - m_i}{\sigma_i} \qquad (3.3)$$

$$(i = 1, 2, \dots, k, l = 1, 2, \dots, n)$$

After normalization, the mean and standard deviation of a variable become 0 and 1, respectively. The correlation coefficient between the ith and j^{th} variables before normalization is given by the following equation:

$$r_{ij} = \frac{1}{n}\left[\frac{X_{il} - m_i}{\sigma_i} \times \frac{X_{jl} - m_j}{\sigma_j} + \dots + \frac{X_{in} - m_i}{\sigma_i} \times \frac{X_{jn} - m_j}{\sigma_j}\right] \qquad (3.4)$$

The same coefficient after normalization is given by the following equation:

$$r_{ij} = \frac{1}{n}(x_{i1}x_{j1} + \dots + x_{in}x_{jn}) \qquad (3.5)$$

Using the correlation coefficients, (r_{ij}) inverse matrix $A = (a_{ij})$ is calculated as the database that is called the Mahalanobis Space. The inverse matrix, A, is given by the following equation:

$$A = (a_{ij}) = \begin{bmatrix} r_{11} & r_{12} & \dots & r_{1k} \\ r_{21} & r_{22} & \dots & r_{2k} \\ \cdot & & & \\ \cdot & & & \\ \cdot & & & \\ r_{k1} & r_{k2} & \dots & r_{kk} \end{bmatrix}^{-1} = \begin{bmatrix} a_{11} & a_{12} & \dots & a_{1k} \\ a_{21} & a_{22} & \dots & a_{2k} \\ \cdot & & & \\ \cdot & & & \\ \cdot & & & \\ a_{k1} & a_{k2} & \dots & a_{kk} \end{bmatrix} \qquad (3.6)$$

To summarize, a database (a_{ij}) $(i = 1, 2, \dots, k, j = 1, 2, \dots, k)$ is calculated by using m_i and σ_i from n sets of data.

3.4 MAHALANOBIS DISTANCE

Let X_1, X_2, ..., and X_k be the surveyed results of *k* items from one person. From these results, m_i and s_i are calculated. From these values, a Mahalanobis Distance or normalized variance (denoted by D^2) is determined as follows:

$$D^2 = \frac{1}{k}\sum_{ij=1}^{k} a_{ij}\frac{X_i - m_i}{\sigma_i} \times \frac{X_j - m_j}{\sigma_j}$$

D^2 has the following properties: When *n* pieces of D^2 are calculated from *n* sets of a normalized space, their average is equal to 1. Mahalanobis approximated the distribution of D^2 to an F distribution with the degrees of freedom of *k* for its numerator and for its denominator. This assumption is appropriate when there is a large number of items (variables). What is important, however, is not the rigorousness of the distribution but the fact that we need to forecast the unpaid amounts. For this purpose, the SN ratio in Taguchi methods is used to evaluate the error(s) of forecasting. The Mahalanobis Space determined from the normalized space only gives a base point and a scale for judgment.

3.5 ESTIMATION OF THE UNPAID AMOUNT AND ITS ESTIMATING ERROR BY USING THE SN RATIO

In order to verify that Mahalanobis Distance is useful for judgments, you should calculate the Mahalanobis Distances of some people who did not pay their dues.

Let the total unpaid amount at the time some people's credit cards were terminated be *M*, and let the unpaid amounts (the losses to the company) of each person be M_1, M_2, ..., and M_l. The Mahalanobis Distance of these people should be denoted as D_1^2, D_2^2, ..., and D_l^2. Refer to Table 3-2.

No one has conducted a study yet to suggest whether D^2 or the square root of D^2 should be used as the output. But we suggest using the square root:

TABLE 3-2 Unpaid Amount and the Square Root of Mahalanobis Distance (M: UNPAID AMOUNT, $y = \sqrt{D^2}$)

Signal	M_1	M_2	...	M_l
Output	y_1	y_2	...	y_l

$$y_i = \sqrt{D_i^2} \quad (i = 1, 2, \ldots, l) \tag{3.7}$$

The following calculations are used to decompose the total variation into the variation of proportional term, denoted by S_β, and the error variation, denoted by S_e. The calculations are identical to those for the SN ratio in Taguchi methods.

Total Variation: $S_T = y_1^2 + y_2^2 + \ldots y_l^2$

(degrees of freedom $f = l$) (3.8)

Proportional Term:

$$S_\beta = \frac{1}{r}(M_1 y_1 + M_2 y_2 + \ldots + M_l y_l)^2 \tag{3.9}$$

$(f = l)$

Magnitude of Input: $r = M_1^2 + M_2^2 + \ldots + M_l^2$ (3.10)

Total Variation: $S_e = S_T - S_\beta$ (3.11)

$(f = l - 1)$

From these results, we obtain an SN ratio. The SN ratio, denoted by η, is the reciprocal of the forecasting error on the decibel scale.

$$\eta = 10\log\frac{\frac{1}{r}(S_\beta - V_e)}{V_e} \tag{3.12}$$

where $1/r\,(S_\beta - V_e)$ is the estimate of β^2 and the error variance, and the estimate of σ^2 is calculated by the following equations:

$$V_e = \frac{S_e}{l - 1} \tag{3.13}$$

$$y = \beta M \tag{3.14}$$

$$\beta = \frac{1}{r}(M_1 y_1 + M_2 y_2 + \ldots + M_l y_l) \tag{3.15}$$

The unpaid amount of a customer is estimated as follows:

$$M = \frac{D}{\beta} \pm 3\frac{1}{\sqrt{\eta}}$$

$$\left(\eta \text{ in anti-log} \rightarrow \frac{\beta^2}{\sigma^2}\right) \tag{3.16}$$

$$\text{where } y = \sqrt{D^2}$$

A financial institution determines the line of credit based on the Mahalanobis Distance of a customer. This process is called tuning and is used in design specification.

3.6 RATIONALIZATION OF ITEM SELECTION

As described previously, there are k items selected in order to construct a Mahalanobis Space. To improve the efficiency of forecasting, the unimportant items should be removed. To detect these unimportant items, consider that all items are Level-2 control factors and are assigned to a Level-2 series orthogonal array. If there are 50 items, an L_{64} orthogonal array is used. In general, let the orthogonal array L_N be used. The two levels of an item are set as follows:

- Level 1: Do not use the item to construct the space.
- Level 2: Use the item to construct the space.

Thus, there will be n different combinations of certain items (Level 2) but not the other items (Level 1). As a result, each of the n combinations has a different evaluating capability. To determine which item is important and which is not, an SN ratio is calculated by using Equation (3.12) for each of the n combinations. From these SN ratios, the effect of each item assigned to each row is calculated; then, the optimum combination is determined.

3.7 DESIGN OF A BUSINESS SYSTEM

The two key tasks in a business system are the judgment from the collected information and actions that follow the judgment. How to collect information is most important for these decisions, and the person who is in charge has this responsibility. Once the information is collected, the selection of a judging method and the estimation of its precision are essential. In medical science, these tasks would involve the selection of a diagnosis method and the estimation of its precision. In the case of geophysics, we would select a method to forecast earthquakes or weather conditions and the estimation of the method's precision. This approach summarizes all of the collected information. In quality engineering, these pieces of information are called signals. Because these pieces of information were collected from observations, they are called passive signals.

In order to treat such passive information, regression (linear or non-linear) analysis or theoretical equations have traditionally been used. These approaches have one element in common, however: only one equation for diagnosis or forecasting. The Mahalanobis Distance, in contrast, uses the entire pattern. Construction of Mahalanobis Space and evaluation and improvement through the use of the SN ratio are indeed the universal approaches for passive-type functions. Although there was an illustration of only one example in this chapter, this approach can be applied to various business systems.

SECTION ONE

HEALTH CARE

CHAPTER FOUR

DIAGNOSIS OF A SPECIAL HEALTH CHECK

4.1 DIAGNOSIS OF LIVER FUNCTION

Giving a diagnosis means drawing conclusions by summarizing the information of a group of characteristics with different units or scales. From a quality-engineering viewpoint, diagnosing involves measurement, and measurement requires two values: a base point and a unit. For example, the base point of a weighing scale is zero (when nothing is placed on the scale). The unit is similar to a pound or a kilogram. With these two values, we can measure an object to tell how far the object is from the base point.

When diagnosing health, what we want to know is whether a person is healthy or not. The base point, therefore, is a healthy person. The measuring unit is ideally a quantifiable variable (rather than a digitally expressed variable), such as healthy or not healthy. Defining the base point and the unit is a difficult challenge, however. Because nobody can truly define a healthy person, a group of healthy people are selected as the group to define zero-point and unit distance. How to select the group is determined by the doctor, based on his or her knowledge and decisions. For example, the people who have been tested for certain diseases in the past two or three years and who are still diagnosed as healthy can be defined as the healthy group. Selection of healthy group is a doctor's freedom in an information system design. The objective of a physical checkup is forecasting whether there will be serious disease before the next checkup. Based on the data collected from these people, a Mahalanobis Space will be constructed to be used as a base group to determine zero-point and unit distance. In the case of checking liver function, there are 15 biochemical testing items—part of which are shown in Table 4-1. This application is one of the earliest applications of Mahalanobis Distance in the diagnosis of liver function (published by Tetsuji Kanetaka of Tokyo Postal Hospital).[1]

The objective of the study was not a general health checkup but a special checkup project to discover liver disease from the people who were older than a certain age. Age and gender were added for diagnosis, because the distribution of the 15 items are quite different, depending on age and gender. In other words, a correlation exists. By using the test results from a group of 200 people who were judged to be healthy, we made a calculation to construct a Mahalanobis Space of a healthy group. This space will be used as the base point.

Next, a person who is about to be diagnosed will receive the same tests as shown in Table 4-1. From the test results, the Mahalanobis Distance of this person is calculated by using

TABLE 4-1 Biochemical Test Items and Their Clinical Implication

Test Item	Abbreviation	Normal Values Used in Hospital	Clinical Significance
Total Protein in Blood Serum	TP	6.5-7.5g/dl	Decreases with severe liver disturbance (hepatopathy) and anemia (due to albumin level reduction)
Albumin in Blood Serum	Alb	3.5-4.5g/dl	Decreases with severe liver disturbance (hepatopathy), anemia, and nephrosis (such as cirrhosis and fulminant hepatitis)
A/G Ratio	A/G	1.00-1.80	Similar to albumin (calculated from TP and Alb; Alb/(TP-Alb)
Cholinesterase	ChE	0.60-1.00 dpH	Almost the same as albumin
Glutamate O Transaminase	GOT	2-25 units	Represented in liver cell disorder (increases with hepatitis and poisonous liver disorder)
Glutamate P Transaminase	GPT	0-22 units	Similar to GOT (more specific with liver compared to GOT)
Lactate Dehydrogenase	LHD	130-250 units	Elevated in severe liver distrubance (hepatopathy), blood, heart, and muscle diseases
Alkaline Phosphatase	ALP	2.0-10.0 units	Increases with bile duct disease, cancer, bone diseases
r-Glutamyl Transpeptidase	r-GPT	0-68 units	Increases with bile duct disease and cancer

[1]Kanetaka, Tatsuji. "Diagnosis of a Special Health Check Using Mahalanobis Distance," *ASI Journal*, Vol. 3, No. 1, 1990.

the Mahalanobis Space determined from the 200 people. Table 4-2 shows the test results of 200 healthy people.

First, the raw data are normalized, as shown in Chapter 2, "A Detailed Example of MTS." Then, the correlation coefficient of each pair of items is calculated in order to construct a correlation matrix, as shown in Table 4-3.

Next, the inverse matrix of this correlation matrix is constructed as shown in Table 4-4.

By using this matrix and normalized data, Mahalanobis Distances (D^2) are calculated as shown in Table 4-2.

TABLE 4-2 Test Results and D^2 from a Normal Group

CasN	Age	Sex	TP	Alb	ChE	AST	ALT	LDH	ALP	GTP	LAP	TCh	TG	PL	Cr	BUN	UA	D Square/f
1	48	1	7.2	4.9	495	26	22	110	235	7	67	183	63	196	1.0	17	3.2	0.37270957
2	40	1	7.4	5.0	524	26	20	108	146	7	68	191	55	2.6	1.1	15	3.7	0.40921106
3	41	1	7.7	5.3	461	24	22	108	171	11	64	179	77	196	1.0	13	2.9	0.42452956
4	52	1	7.1	4.8	552	21	18	98	182	8	63	193	81	2.7	0.9	14	2.7	0.52648390
5	41	1	7.7	5.3	524	20	20	96	171	10	63	181	101	198	0.8	14	4.0	0.53650114
6	45	1	7.5	5.2	569	26	24	122	225	16	75	192	66	202	0.8	13	2.9	0.53859066
7	31	1	7.9	5.2	547	21	20	90	153	8	65	165	64	178	0.8	11	3.3	0.57428233
8	55	1	7.3	4.8	530	29	26	118	250	10	73	174	64	186	1.0	13	3.2	0.59251079
9	51	1	7.5	4.8	530	26	20	100	218	8	61	201	98	216	0.9	14	2.7	0.59806885
10	32	1	7.3	5.2	524	20	16	98	203	8	69	170	40	192	0.9	15	3.2	0.60030919
11	35	1	7.4	5.2	501	16	18	92	132	7	70	171	55	186	1.0	12	4.3	0.62633442
12	32	1	7.5	5.3	416	24	20	115	189	12	76	156	64	169	0.9	15	3.4	0.63254621
13	54	1	7.2	5.0	456	33	25	119	210	7	67	168	40	181	0.9	13	2.7	0.64205212
14	31	1	7.4	5.2	399	20	20	114	185	7	71	175	46	191	0.9	14	2.5	0.64498103
15	31	1	7.2	4.9	569	23	22	106	136	7	72	174	63	188	1.0	14	2.6	0.65716589
16	52	1	7.0	5.2	473	20	13	109	164	9	64	160	60	171	0.8	13	2.9	0.67311699
⋮	⋮	⋮	⋮	⋮	⋮	⋮	⋮	⋮	⋮	⋮	⋮	⋮	⋮	⋮	⋮	⋮	⋮	⋮
196	56	10	7.4	5.2	630	31	26	118	257	80	81	188	134	215	1.5	19	3.9	1.73724565
197	57	10	7.9	5.2	678	41	42	119	325	81	90	151	150	167	1.4	14	5.0	1.75533058
198	33	10	8.0	5.2	678	21	27	107	178	17	61	200	190	225	1.4	15	6.0	1.75614022
199	59	10	7.4	5.0	654	20	22	101	225	17	61	206	143	223	1.0	13	4.1	1.92158710
200	57	10	6.9	4.9	562	29	25	88	182	26	78	138	85	178	1.4	18	5.2	2.39194641

TABLE 4-3 Correlation Matrix of 200 Healthy People

	1	2	3	4	5	6	7	8	9
1	1	-0.296785	-0.277785	-0.402941	-0.220432	0.101427	0.040969	0.208286	0.292511
2	-0.296765	1	0.10322	0.415606	0.690461	0.287106	0.379001	-0.107578	-0.047273
3	-0.277785	0.10322	1	0.427084	0.201501	0.083581	0.138772	0.071889	0.011293
4	-0.402941	0.415806	0.427084	1	0.315205	0.037661	0.055853	0.009715	-0.105683
5	-0.220432	0.890461	0.201501	0.315205	1	0.345168	0.385324	0.062584	-0.057415
6	0.101427	0.287105	0.083581	0.037661	0.345168	1	0.790343	0.315633	0.17354
7	0.040969	0.379001	0.138772	0.055853	0.385324	0.790343	1	0.143192	0.068151
8	0.208286	-0.107578	0.071889	0.009715	0.062584	0.315633	0.143192	1	0.228903
9	0.292511	-0.047716	0.011293	-0.105638	-0.057415	0.17654	0.068151	0.228903	1
10	-0.103656	0.647273	0.176563	0.269008	0.578008	0.550078	0.587687	0.128697	0.064717
11	-0.111709	0.394649	0.182088	0.218931	0.42873	0.534703	0.507497	0.227272	0.146902
12	0.26439	-0.237283	0.069506	-0.136104	0.012325	0.14831	0.134228	0.250152	0.171041
13	0.13505	0.269333	0.158055	0.100119	0.366813	0.319548	0.373141	0.102511	0.120639
14	0.283289	-0.222137	0.077988	-0.135149	0.032099	0.167903	0.147887	0.25651	0.187784
15	-0.291931	0.885853	0.150179	0.364531	0.64428	0.321345	0.397622	-0.062997	-0.074507
16	-0.018557	0.25412	-0.119402	0.090886	0.13518	0.181238	0.10816	0.0095115	-0.096074
17	-0.281627	0.797969	0.19783	0.413004	0.675257	0.359093	0.434501	-0.014632	-0.061275

	10	11	12	13	14	15	16	17
1	-0.103556	-0.111769	0.26439	0.13505	0.283289	-0.291931	-0.018557	-0.281627
2	0.394649	-0.237263	0.269333	-0.222137	0.885853	0.885853	0.254112	0.797969
3	0.178583	0.182088	0.069506	0.158055	0.077986	0.150179	-0.119402	0.19783
4	0.269008	0.218931	-0.136104	0.100119	-0.135149	0.364531	0.090886	0.413004
5	0.578008	0.42873	0.012325	0.368813	0.032099	0.64428	0.13518	0.675267
6	0.550078	0.534703	0.14831	0.319546	0.167903	0.321345	0.181238	0.359093
7	0.567687	0.507497	0.134228	0.373141	0.147887	0.397622	0.10916	0.434501
8	0.128697	0.227272	0.250152	0.102511	0.25651	-0.062997	0.095115	-0.014632
9	0.064717	0.146902	0.171041	0.120639	0.167784	-0.074507	-0.096074	-0.061275
10	1	0.682581	0.051788	0.436679	0.079144	0.612022	0.138442	0.649484
11	0.682581	1	0.158948	0.342114	0.15221	0.444857	0.048007	0.465275
12	0.051788	0.158948	1	0.310286	0.96679	-0.140092	-0.004247	-0.023233
13	0.46879	0.342114	0.310286	1	0.318056	0.266723	-0.04053	0.351668
14	0.079144	0.15221	0.96879	0.318056	1	-0.118803	0.024808	-0.011287
15	0.612022	0.44857	-0.140092	0.266723	-0.118803	1	0.278543	0.770913
16	0.138442	0.048007	-0.004247	-0.04053	0.024808	0.278543	1	0.178618
17	0.649484	0.465275	-0.023233	0.351868	-0.011287	0.770913	0.178618	1

TABLE 4-4 Inverse Matrix of 200 Healthy People

	1	2	3	4	5	6	7	8	9	10	11	12	13	14	15	16	17
1	1.591597	-0.00289	0.307009	0.297082	0.117747	-0.08763	-0.11603	-0.19282	-0.3041	-0.11325	0.247721	0.337266	-0.28384	-0.55198	0.148263	-0.02768	0.197724
2	0.00289	8.135734	0.657994	-0.70552	-1.28118	0.626837	-0.4388	0.379145	-0.57621	-1.48183	0.748064	-0.19245	-0.07682	1.358014	-4.27704	-0.31639	-1.52538
3	0.307009	0.657994	1.441935	-0.59418	-0.16918	0.135761	-0.25838	-0.06581	-0.12295	-0.11518	0.07014	0.222736	-0.09671	-0.3038	-0.31453	0.191141	-0.02313
4	0.297082	-0.70552	-0.59418	1.67893	0.101376	0.008985	0.271992	-0.14324	0.088065	0.071164	-0.15657	0.02592	-0.04919	0.054758	0.317199	-0.10313	-0.296112
5	0.117747	-1.28118	-0.16918	0.101376	2.357242	-0.1972	0.10984	-0.19344	0.20004	-0.03421	-0.12141	0.210494	-0.23545	-0.43956	0.077441	0.108166	-0.42949
6	-0.08163	0.626837	0.135761	0.008986	-0.1972	3.403374	-2.26612	-0.48309	-0.29733	-0.43649	-0.34811	0.331809	0.04377	-0.15597	-0.10795	-0.33812	-0.10442
7	-0.11603	-0.4388	-0.25838	0.271992	0.10984	-2.26612	3.192262	0.274867	0.252132	-0.17186	-0.13304	-0.24049	-0.1948	0.10623	-0.00897	0.147444	-0.15322
8	-0.19282	0.379145	-0.06581	-0.14324	-0.19344	-0.48309	0.274867	1.337763	-0.15743	-0.05584	-0.17889	-0.10274	0.063683	-0.02842	0.022238	-0.14265	0.012239
9	-0.3041	-0.57621	-0.12295	0.088065	0.20004	-0.29733	0.252132	-0.15743	1.246912	0.100554	-0.21799	-0.11838	-0.03411	-0.00637	0.239915	0.157056	0.130938
10	-0.11325	-1.48183	-0.11518	0.071164	-0.03421	-0.43649	-0.17186	-0.05564	0.100554	3.320745	-1.24712	0.928091	-0.33528	-1.004	0.386188	0.041256	-0.34958
11	0.247721	0.748064	0.07014	-0.15857	-0.12141	-0.34811	-0.13304	-0.17889	-0.21799	-1.24712	2.301979	-0.88035	-0.00132	0.753881	-0.63704	0.150582	-0.0356
12	0.337286	-0.19245	0.222736	0.02592	0.210493	0.331808	-0.24049	-0.10274	-0.11838	0.928092	-0.88035	16.23363	-0.29335	-15.6144	0.588934	0.273758	-0.38324
13	-0.28384	-0.07682	-0.09671	-0.04919	-0.23545	0.04377	-0.1948	0.063683	-0.03411	-0.33528	-0.00132	-0.29335	1.536702	-0.09646	0.042757	0.16857	-0.14475
14	-0.55198	1.358014	-0.3038	0.054758	-0.43956	-0.15597	0.10623	-0.02842	-0.00637	-1.004	0.753881	-15.6144	-0.09645	16.52648	-0.82594	-0.46251	-0.01833
15	0.146263	-1.27704	-0.31453	0.137199	0.077441	-0.10795	-0.00897	0.022238	0.239915	0.386188	-0.63704	0.588934	0.042757	-0.82594	5.415279	-0.33033	-0.69072
16	-0.02766	-0.31639	0.194414	-0.10313	0.108166	-0.33812	0.147444	-0.14265	0.157056	0.041255	0.150582	0.272759	0.16657	-0.46251	-0.33033	1.248879	0.120447
17	0.197724	-1.52538	-0.02313	-0.29612	-0.42949	-0.10442	-0.15322	0.012239	0.130938	-0.34956	-0.0356	-0.36324	-0.14475	-0.01883	-0.69072	0.120447	3.599122

The average of the Mahalanobis Distance of a healthy group is equal to 1. When the decibel (10 times logarithm) scale is used, the average is equal to zero. But the patients who have liver disease will show much larger values. The more severe the disease, the bigger the distance. On the other hand, the more the patient recovers from the illness, the smaller the distance becomes.

In this study, from the 200 people who were judged as healthy from the test, a Mahalanobis Space was constructed. Next, a new group of 95 people were selected to receive the same tests.

Traditionally, the result of each item in the table is compared with the range of normal values in Table 4-1. If any result is beyond the normal range, the person is instructed to receive a complete checkup. Table 4-5 shows the results of judgment based on the traditional approach.

Based on the results of the complete checkup, the people were classified into the following four groups:

- Group 1 No liver disease
- Group 2 No liver disease, but did not follow the doctor's instructions (such as diet or alcoholic intake) before the test
- Group 3 Slight disease
- Group 4 Significant disease

According to the traditional method, 39 people out of 66 were misjudged abnormal. The rate of mistake is 59 percent, which is considerably high.

Tables 4-6a, b, c, and d show the test results and Mahalanobis Distances of the four groups.

From the test results of each person, a Mahalanobis Distance was calculated by using the Mahalanobis Space.

TABLE 4-5 Results of Diagnosis by Using the Traditional Method

GROUP	NORMAL	ABNORMAL	TOTAL
(1)	27	39	66
(2)	1	12	13
(3)	1	10	11
(4)	0	5	5
Total	29	66	95

TABLE 4-6a Test Results and D^2 of New 95 People (Normal Group)

CasN	Age	Sex	TP	Alb	ChE	AST	ALT	LDH	ALP	GTP	LAP	TCh	TG	PL	Cr	BUN	UA	FinDg	D Sq./f
1	51	1	7.6	5.2	489	21	21	120	217	13	66	180	112	199	1.1	14	2.8	WNL	0.650870
2	52	1	7.8	4.8	420	26	24	109	232	12	66	180	97	190	1.0	16	2.9	WNL	0.825876
3	45	1	7.8	5.0	357	24	20	119	245	14	62	192	96	209	1.1	12	3.3	WNL	0.878916
4	49	1	7.7	4.8	466	20	16	85	224	14	60	186	61	208	1.0	14	3.0	WNL	0.918397
5	51	1	7.6	4.8	357	24	26	95	222	13	62	207	55	215	1.0	14	3.5	WNL	0.968393
6	49	1	7.6	4.7	450	23	21	90	149	14	53	207	100	222	0.9	14	3.6	WNL	0.999225
7	51	1	7.4	4.9	434	25	22	114	188	12	67	185	102	217	0.9	16	3.2	WNL	1.037695
8	49	1	7.6	5.0	419	26	27	137	193	15	70	212	44	228	1.1	20	3.5	WNL	1.075267
9	52	1	7.8	4.9	370	20	14	103	219	13	69	180	63	205	1.1	14	2.9	WNL	1.136795
10	54	1	7.9	5.2	341	28	28	119	211	14	66	220	106	241	1.1	14	4.4	WNL	1.326849
11	58	1	8.1	4.9	481	22	16	88	209	13	64	202	123	222	1.0	15	2.3	WNL	1.328718
12	41	1	7.4	4.8	295	22	23	90	217	13	56	201	43	212	0.8	15	2.5	WNL	1.367741
13	45	1	7.9	5.2	349	23	30	105	198	14	67	153	69	174	0.8	11	2.7	WNL	1.390987
14	57	1	7.8	4.7	411	22	21	101	241	14	68	210	92	218	0.8	17	3.5	WNL	1.453555
15	52	1	7.5	4.7	322	22	23	107	209	14	64	182	94	195	1.2	13	3.7	WNL	1.489989
16	52	1	7.6	4.8	394	23	23	127	196	13	78	178	108	182	0.9	16	2.6	WNL	1.501135
17	59	1	7.6	4.8	302	26	21	111	247	13	66	175	64	181	0.8	15	3.0	WNL	1.517162
18	53	1	7.9	5.2	371	22	14	108	279	14	64	198	101	229	0.9	14	2.9	WNL	1.554667
19	54	1	7.7	4.8	302	25	20	125	230	13	61	207	98	213	0.9	15	2.8	WNL	1.619822
20	51	1	7.5	4.6	388	22	23	121	205	12	61	196	112	205	1.1	11	3.1	WNL	1.635945
21	48	1	7.5	4.8	434	21	19	133	178	12	68	168	81	197	1.1	11	3.0	WNL	1.719802
22	56	1	7.5	5.0	295	23	23	100	227	15	64	220	66	217	0.8	14	2.8	WNL	1.722244
23	42	1	7.6	4.8	458	23	20	122	209	13	58	187	58	181	0.9	14	2.7	WNL	1.723828
24	47	1	7.5	4.8	326	22	19	114	310	14	62	186	26	207	0.8	18	2.5	WNL	1.772167
25	52	1	7.8	5.2	365	28	35	98	302	20	67	234	176	244	1.1	14	2.9	WNL	1.859646
26	50	1	7.6	4.4	388	22	25	93	172	11	60	187	70	202	0.8	14	2.5	WNL	1.894197
27	41	1	7.6	4.8	295	18	16	124	175	11	62	182	96	191	0.9	14	2.6	WNL	1.920668
28	28	1	7.5	4.8	429	18	14	107	263	13	63	189	92	2217	1.2	10	3.0	WNL	1.947224
29	41	1	7.6	4.9	372	22	18	100	154	10	58	226	183	232	0.9	9	3.1	WNL	1.979011

TABLE 4-6b Test Results and D^2 of New 95 People (Normal Group, but Did Not Follow Doctor's Instructions)

CasN	Age	Sex	TP	Alb	ChE	AST	ALT	LDH	ALP	GTP	LAP	TCh	TG	PL	Cr	BUN	UA	FinDg	D Sq./f
65	35	10	8.3	4.8	424	25	21	113	225	32	61	181	87	197	1.5	14	6.0	WNL	2.837303
66	56	10	8.1	5.0	473	22	30	88	198	92	77	198	82	212	1.4	17	5.2	WNL	3.391276
67	52	1	7.8	5.2	365	32	31	98	302	20	67	234	198	239	1.1	14	2.9	SAB	2.207052
68	49	1	7.7	4.8	466	36	41	138	224	14	60	248	198	262	1.0	14	3.0	SAB	2.349045
69	53	1	8.1	5.0	318	24	31	110	217	24	79	240	208	257	0.9	14	4.1	SAB	2.583183
70	48	1	6.3	4.4	442	25	23	137	222	10	61	230	155	255	0.9	14	2.7	SAB	3.480291
71	43	1	7.8	4.9	365	22	21	148	232	14	70	237	194	232	0.8	13	3.5	SAB	3.486126
72	57	1	7.6	5.2	341	25	18	108	246	16	76	200	132	245	1.0	17	2.8	SAB	3.537835
73	43	10	7.6	5.0	521	32	34	109	243	96	85	223	178	242	1.0	14	6.0	SAB	3.154030
74	47	10	7.5	5.0	563	31	35	113	269	92	74	233	166	242	1.0	14	6.5	SAB	3.170688
75	44	10	7.7	4.9	546	31	38	111	154	87	68	198	176	232	1.3	14	6.1	SAB	3.201982
76	57	10	7.8	4.8	578	28	32	117	232	102	60	212	175	244	1.2	16	4.7	SAB	4.617025
77	54	10	7.9	4.9	546	34	36	118	286	104	60	217	169	254	1.2	18	5.2	SAB	5.161107
78	42	10	7.9	5.0	455	30	38	89	232	108	71	210	163	257	1.2	19	4.5	SAB	6.737166

TABLE 4-6c Test Results and D^2 of New 95 People (Slight Disease)

CasN	Age	Sex	TP	Alb	ChE	AST	ALT	LDH	ALP	GTP	LAP	TCh	TG	PL	Cr	BUN	UA	FinDg	D Sq./f
79	52	1	7.9	5.0	237	31	33	145	261	14	67	273	292	294	1.4	18	4.2	SLI	6.989781
80	56	1	6.8	4.7	151	38	45	111	358	58	92	198	112	251	0.9	14	2.9	SLI	8.422783
81	47	1	6.9	4.8	182	24	18	134	250	13	68	183	189	230	1.9	15	3.7	SLI	10.086479
82	48	10	8.3	5.0	360	34	43	135	250	58	71	234	318	259	1.5	14	5.2	SLI	7.193340
83	47	10	7.6	5.0	277	22	25	77	321	121	95	159	171	217	1.3	11	5.6	SLI	10.815891
84	46	10	7.4	5.2	318	46	88	121	256	44	81	235	151	268	1.7	14	7.0	SLI	10.985479
85	46	10	7.4	5.2	318	46	88	121	256	44	81	235	151	286	1.7	14	7.0	SLI	13.841776
86	55	10	7.4	4.9	273	27	38	116	198	44	72	237	419	289	1.5	17	6.4	SLI	15.067843
87	50	10	7.0	5.0	290	30	55	127	204	56	84	323	416	303	1.7	13	7.6	SLI	16.128505
88	56	10	8.1	5.8	261	25	28	134	204	80	82	304	188	378	1.5	16	5.7	SLI	19.408344

TABLE 4-6d Test Results and D^2 of New 95 People (Significant Disease)

CasN	Age	Sex	TP	Alb	ChE	AST	ALT	LDH	ALP	GTP	LAP	TCh	TG	PL	Cr	BUN	UA	FinDg	D Sq./f
89	41	1	7.6	4.9	108	27	31	167	354	123	117	279	176	273	1.2	15	2.9	LVD	11.793325
90	51	10	7.2	5.2	417	38	45	134	305	181	99	230	182	220	2.0	16	7.4	LVD	12.129345
91	68	10	7.6	5.0	273	35	46	100	214	240	93	221	185	257	1.5	16	4.0	LVD	19.921832
92	56	10	6.5	4.8	354	38	40	106	250	147	83	132	424	263	1.3	16	6.6	LVD	42.752055
93	47	10	6.7	4.8	174	144	171	136	222	68	73	110	364	229	1.5	14	6.6	LVD	77.149165
94	51	10	5.4	3.6	46	149	168	212	211	118	78	80	105	220	1.4	13	6.9	LVD	96.048353
95	56	10	6.1	3.8	45	135	93	447	315	227	121	128	356	268	1.1	12	6.7	LVD	132.381753

The Mahalanobis Distances of Group 2 are higher than the distances of Group 1, because the people in Group 2 did not follow the doctor's instructions. These people have no disease, however. If the results of Group 2 are not compared, it is obvious from Table 4-6 that with a threshold value of, say, 4 or 5, there is no mistake in diagnosis.

4.2 SELECTION OF CHARACTERISTICS

In this case, 17 characteristics were used for pattern recognition. Some of the characteristics might not contribute to the function of pattern recognition. To screen those non-contributing characteristics, an orthogonal array is used to assign all characteristics and collect the people from both the healthy and non-healthy groups, in order to assign their D^2 outside the orthogonal array as the signal factor. Because there are 17 characteristics, an orthogonal array L_{32} is used. Tables 4-7a and 4-7b show the test results of eight people who were arbitrarily picked from Table 4-2 and eight non-healthy people who were also arbitrarily picked from Table 4-6c. Table 4-8 shows the D^2 of the two groups. Table 4-9 shows the layout and calculated SN ratios by using a zero-point proportional equation.

TABLE 4-7a–b Test Results of Healthy People and Non-Healthy People

Table 4.7: Test Results of Healthy People

S. No.	X1	X2	X3	X4	X5	X6	X7	X8	X9	X10	X11	X12	X13	X14	X15	X16	X17
1	48	1	7.2	4.9	495	26	22	110	235	7	67	183	63	196	1	17	3.2
2	40	1	7.4	5	524	26	20	108	146	7	68	191	55	206	1.1	15	3.7
3	41	1	7.7	5.3	461	24	22	108	171	11	64	179	77	196	1	13	2.9
4	52	1	7.1	4.8	552	21	18	98	182	8	63	193	81	207	0.9	14	2.7
5	41	1	7.7	5.3	524	20	20	96	171	10	63	181	101	198	0.8	14	4
6	45	1	7.5	5.2	569	56	4	122	225	16	75	192	66	202	0.8	13	2.9
7	31	1	7.9	5.2	547	21	20	90	163	8	65	165	64	178	0.8	11	3.3
8	55	1	7.3	4.8	530	29	26	118	250	10	73	174	64	186	1	13	3.2

Table 4.7: Test Results of Non-Healthy People

S. No.	X1	X2	X3	X4	X5	X6	X7	X8	X9	X10	X11	X12	X13	X14	X15	X16	X17
1	52	1	7.9	5	237	31	33	145	261	14	67	273	292	294	1.4	18	4.2
2	56	1	6.8	4.7	151	38	45	111	358	58	92	198	112	251	0.9	14	2.9
3	47	1	6.9	4.8	182	24	18	134	250	13	68	183	189	230	1.9	15	3.7
4	48	10	8.3	5	360	34	43	135	250	58	71	234	318	259	1.5	14	5.2
5	47	10	7.6	5	277	22	25	77	321	121	95	159	171	217	1.3	11	5.6
6	46	10	7.4	5.2	318	46	88	121	256	44	81	235	151	268	1.7	14	7
7	46	10	7.4	5.2	318	46	88	121	256	44	81	235	151	286	1.7	14	7
8	55	10	7.4	4.9	273	27	38	116	198	44	72	237	419	289	1.5	17	6.4

TABLE 4-8 D^2 of Healthy and Non-Healthy People

	1	2	3	4	5	6	7	8
Healthy	0.378374	0.431373	0.403562	0.500211	0.515396	0.495501	0.583142	0.565654
Non-Healthy	7.727406	8.416294	10.29148	7.205157	10.59075	10.55711	13.31775	14.81278

TABLE 4-9 Layout and SN Ratio

	X1	X2	X3	X4	X5	X6	X7	X8	X9	X10	X11	X12	X13	X14	X15	X16	X17	S/N Ratio
	1	2	3	4	5	6	7	8	9	10	11	12	13	14	15	16	17	
1	1	1	1	1	1	1	1	1	1	1	1	1	1	1	1	1	1	-7.1490
2	1	1	1	1	1	1	1	1	1	1	1	1	1	1	1	2	2	-7.0375
3	1	1	1	1	1	1	1	2	2	2	2	2	2	2	2	1	1	-10.5163
4	1	1	1	1	1	1	1	2	2	2	2	2	2	2	2	2	2	-10.6545
5	1	1	1	2	2	2	2	1	1	1	1	2	2	2	2	1	1	-10.8970
6	1	1	1	2	2	2	2	1	1	1	1	2	2	2	2	2	2	-11.0383
7	1	1	1	2	2	2	2	2	2	2	2	1	1	1	1	1	1	-8.6724
8	1	1	1	2	2	2	2	2	2	2	2	1	1	1	1	2	2	-8.8603
9	1	2	2	1	1	2	2	1	1	2	2	1	1	2	2	1	1	-9.8028
10	1	2	2	1	1	2	2	1	1	2	2	1	1	2	2	2	2	-10.1635
11	1	2	2	1	1	2	2	2	2	1	1	2	2	1	1	1	1	-4.7706
12	1	2	2	1	1	2	2	2	2	1	1	2	2	1	1	2	2	-4.8232
13	1	2	2	2	2	1	1	1	1	2	2	2	2	1	1	1	1	-11.3041
14	1	2	2	2	2	1	1	1	1	2	2	2	2	1	1	2	2	-11.4861
15	1	2	2	2	2	1	1	2	2	1	1	1	1	2	2	1	1	-7.3078
16	1	2	2	2	2	1	1	2	2	1	1	1	1	2	2	2	2	-7.9662
17	2	1	2	1	2	1	2	1	2	1	2	1	2	1	2	1	2	-8.6071
18	2	1	2	1	2	1	2	1	2	1	2	1	2	1	2	2	1	-8.3086
19	2	1	2	1	2	1	2	2	1	2	1	2	1	2	1	1	2	-10.586
20	2	1	2	1	2	1	2	2	1	2	1	2	1	2	1	2	1	-10.3454
21	2	1	2	2	1	2	1	1	2	1	2	2	1	2	1	1	2	-6.4632
22	2	1	2	2	1	2	1	1	2	1	2	2	1	2	1	2	1	-6.3633
23	2	1	2	2	1	2	1	2	1	2	1	1	2	1	2	1	2	-8.3563
24	2	1	2	2	1	2	1	2	1	2	1	1	2	1	2	2	1	-8.2601
25	2	2	1	1	2	2	1	1	2	2	1	1	2	2	1	1	2	-11.7289
26	2	2	1	1	2	2	1	1	2	2	1	1	2	2	1	2	1	-11.4254
27	2	2	1	1	2	2	1	2	1	1	2	2	1	1	2	1	2	-6.8132
28	2	2	1	1	2	2	1	2	1	1	2	2	1	1	2	2	1	-6.0714
29	2	2	1	2	1	1	2	1	2	2	1	2	1	1	2	1	2	-9.7577
30	2	2	1	2	1	1	2	1	2	2	1	2	1	1	2	2	1	-9.5375
31	2	2	1	2	1	1	2	2	1	1	2	1	2	2	1	1	2	-6.568
32	2	2	1	2	1	1	2	2	1	1	2	1	2	2	1	2	1	-6.3665

In the array, levels were assigned as follows:

- Level 1: used
- Level 2: not used

Table 4-10 is the response table constructed from the SN ratios in Table 4-9.

In the table, there are eight characteristics (such as x1, x2, x3, x6, and ...) that negatively contributed to recognition. In other words, using these characteristics resulted in a lower SN ratio. After removing these eight characteristics and using the remaining characteristics, the D^2 of both healthy and non-healthy people were calculated as shown in Table 4-11.

Obviously, the difference in D^2 values between healthy and non-healthy people in Table 4-11 is much greater than the differences in Table 4-8, suggesting a much better pattern-recognizing power.

TABLE 4-10 Response Table

	Averages		
	Used	Not Used	
	Level 1	Level 2	
X1	-8.901224	-8.478037	-0.4232
X2	-8.885956	-8.493306	-0.3926
X3	-8.941493	-8.437768	-0.5042
X4	-8.680637	-8.698424	0.0176
X5	-7.911875	-9.467387	1.5555
X6	-8.974268	-8.404993	-0.5693
X7	-8.681456	-8.697805	0.0163
X8	-9.4475	-7.931763	-1.5157
X9	-8.890325	-8.488937	-0.4014
X10	-7.290057	-10.08921	2.7992
X11	-8.811682	-8.567581	-0.2441
X12	-8.540025	-8.839238	0.2992
X13	-8.3042	-9.075062	0.7709
X14	-8.117194	-9.262068	1.1449
X15	-8.369993	-9.00927	0.6393
X16	-8.7044	-8.674863	-0.0295
X17	-8.572388	-8.806875	0.2345

TABLE 4-11 D^2 of Healthy and Non-Healthy People After Optimization

Confirmation Run								
Healthy	0.33244	0.369519	0.540385	0.784977	0.679767	0.764885	0.497972	0.591484
Non-Healthy	12.3949	13.57104	15.70749	9.646029	17.29453	16.45924	21.14967	24.83029

4.3 THE LOSS FUNCTION OF HEALTH CHECKUPS (DETERMINATION OF THRESHOLD)

You should note that the threshold value must be determined by the use of the loss function, instead of balancing the two types of errors. The loss function is determined in the same way as tolerance: determining the shipping tolerance based on the functional limit.

The threshold value used to determine whether a person should take a complete examination, denoted by D, is calculated as follows:

Let A, A_o, and D* be defined by the following:

A = The cost of taking a complete examination (includes pain or taking a day off of work)

A_o = The loss caused by not taking a complete examination and having the disease show up in the next checkup, or the loss increase after having subjective symptoms followed by taking a complete examination

D* = Mid-value of the Mahalanobis Distance of a patient group that has subjective symptoms

The threshold value to take a complete examination, D, is calculated as follows:

$$D = \sqrt{\frac{A}{A_o}} \times D^*$$

This value is different from disease to disease, because in each case, the values of A, A_o, and D* are different.

CHAPTER FIVE

APPLICATION FOR MEDICAL TREATMENT

5.1 PROBLEMS IN CLINICAL RESEARCH

So-called double-blind tests have been used in the study of clinical treatments. In this method, numerous subjects must participate in the study, which takes a long time and results in a high cost. Because the objective of quality engineering is to improve the efficiency of research works, the same principle can be used for the study of clinical treatments.

Two reasons exist as to why such a large number of patients are required for this type of study. First, there is a large individual difference between patients. Second, in most cases the type of data that is being collected belongs to classified attributes, such as significant effect, effective, or no effect. If it were possible to use continuous variables, which indicate the condition of the patient or the effect of a treatment, the study could be conducted by just observing one or two patients. This approach can be applied to the studies of animals or medicines.

5.2 MAHALANOBIS DISTANCE AND TREATMENT EFFECT

In order to improve the efficiency of a study, you must quantify the status of a patient instead of using digital-type data.

Table 5-1 shows the Mahalanobis Distance of acute virus liver-disease patients by using the characteristics suggested by

Kanetaka.[1] The table shows the Mahalanobis distance of 36 patients when the treatment started, mid term of treatment and the time leaving the hospital. The duration of treatment of individual patient is different, but it is within several weeks.

T_1 Before treatment (entered the hospital)

T_2 Midterm of treatment

T_3 Treatment ended (left the hospital)

From the table, we can see that the Mahalanobis Distances before treatment were in the range of three to five digits, which is large compared to normal people whose average Mahalanobis Distance is equal to 1. With such a large variation, this distribution has no meaning. You should be able to observe the change of the condition of a patient.

The Mahalanobis Distances calculated in Table 5-1 were calculated from the Mahalanobis Space that was constructed from studying 200 healthy people. The range of the distances from these 200 people was from 3.8 to 0.4, while some of the Mahalanobis Distances in T_3 in the table exceed this range. The reason is probably because some of the patients were not completely cured, or their liver was not strong enough yet. A significant reduction exists, however, compared to the distance before treatment.

5.3 THE STUDY USING ONE PATIENT

In the study, you should evaluate the transition of individual patients. The transition is approximated by the following equation:

$$D^2 = D_0^{\,2}\, e^{-\beta T} \qquad (5.1)$$

[1]Taguchi, Genichi. "Application of Mahalanobis Distance for Medical Treatment." Journal of Quality Engineering Forum, Vol. 2, No. 6.

TABLE 5-1 Mahalanobis Distance of Acute Virus Liver Disease Patients

Patients	T_1	T_2	T_3
1	1158.8	11.0	7.4
2	1188.5	51.4	4.9
3	512.9	12.6	15.8
4	1438.8	57.8	3.8
5	15.10.1	16.6	2.1
6	3411.9	18.8	1.9
7	933.3	11.0	4.6
8	5021.1	25.1	3.9
9	3758.3	11.0	2.2
10	602.6	16.6	3.1
11	21677.0	19.6	2.1
12	4897.8	20.4	2.7
13	1088.9	8.6	4.6
14	108.9	6.9	6.2
15	1510.1	18.6	2.9
16	50003.4	39.9	2.8
17	6338.7	21.1	2.7
18	959.4	35.6	16.6
19	1291.2	11.7	2.1
20	7691.3	9.9	11.9
21	1315.2	18.6	3.2
22	1737.8	27.2	7.4
23	31188.9	28.8	3.9
24	8953.6	7.7	6.3
25	3741.1	18.5	4.1
26	21478.3	63.4	3.3
27	2167.7	21.1	2.7
28	30338.9	29.0	1.7
29	4497.8	35.2	2.7
30	19319.6	16.8	5.5
31	28973.4	21.0	15.7
32	3758.4	50.4	3.5
33	8472.2	8.6	1.9
34	16368.2	64.3	11.2
35	5176.1	10.7	3.3
36	13868.0	27.0	17.2

where

D^2 = Mahalanobis Distance after *T* days

D_0^2 = Mahalanobis Distance before treatment

T = Number of days (or hours) treated

β = constant

The equation is converted into a natural logarithm (or common logarithm) scale.

$$\ln D^2 = \ln D_0^2 - \beta T \tag{5.2}$$

$$\ln \frac{D_0^2}{D^2} = \beta T \tag{5.3}$$

letting *y* be

$$y = \ln \frac{D_0^2}{D^2} \tag{5.4}$$

Equation (5.1) is written as

$$y = \beta T \tag{5.5}$$

Let the initial value of Mahalanobis Distance be D_0^2 and the Mahalanobis Distances after T_1, T_2, . . . , and T_k days be D_1^2, D_2^2, . . . , and D_k^2. These values are shown in Table 5-2. Equation (5.7) shows the linear relationship between *T* and *y*. Letting the number of units of linear equation *L* be *r*,

$$y_i = \ln \frac{D_0^2}{D^2} \; (i = 1,2,...,k) \tag{5.6}$$

$$L = 0 \times 0 + T_1 \times y_1 + T_2 \times y_2 + \ldots + T_k \times y_k \tag{5.7}$$

$$r = 0^2 + T_1^2 + T_2^2 + \ldots + T_k^2 \tag{5.8}$$

Constant β in Equation (5.1) is calculated as follows:

$$\beta = \frac{L}{r} \tag{5.9}$$

TABLE 5-2 Time of Treatment and Results

Time (Day)	$T_0 = O$	T_1	T_2	....	T_k	Linear Equation
M.D.	D_0^2	D_1^2	D_2^2	....	D_k^2	
Transformed Results	O	y_1	y_2	....	y_k	L

The SN ratio of treatment is calculated as follows:

$$\eta = 10\log \frac{\frac{1}{r}(S_\beta - V_e)}{V_e} \tag{5.10}$$

where

$$S_\beta = \frac{L^2}{r} \tag{5.11}$$

$$V_e = \frac{1}{k}(S_T - S_\beta) \tag{5.12}$$

$$S_T = 0^2 + y_1^2 + y_2^2 + \ldots + y_k^2 \tag{5.13}$$

The progress of treatment is judged by the β in Equation (5.9). The larger the β, the better the progress. A negative β indicates a change for the worse.

For the individual patient, you must observe whether Mahalanobis Distance D^2 is getting closer to 1 or whether the value is moving the other way. Equation (5.14) is recommended to use for observation, because the value of D^2 is too large in some cases. Or, the square root of D^2 can be used.

$$y = 10\log D^2 \tag{5.14}$$

To forecast time when the patient recovers, the following equation is used (only for the case when the β value is positive):

$$T = \frac{1}{\beta}\ln D_0^2 \tag{5.15}$$

This equation is rewritten from Equation (5.1) by setting D^2 equal to 1. When β is equal to 0 or a negative value, a different treatment might be necessary.

5.4 COMPARISON OF TREATMENT METHODS

To compare two treatment methods, A_1 and A_2, only one patient for each method can be used. Let A_1 be the current method, and let A_2 be the new method. Table 5-3 shows the results of comparing the two treatment methods. In the table, time is denoted by T_0, T_1, . . . , and T_k, where T_0 is the time to start treatment.

The results in Table 5-2 are converted by using the following equation to obtain Table 5-3.

$$y_{ij} = 10\log \frac{D_{io}^{\ 2}}{D_{ij}^{\ 2}} (i = 1,2, j = 1,2,...,k) \tag{5.16}$$

The variations in the table are decomposed as follows (refer to Table 5-4):

TABLE 5-3 Comparison of Two Treatment Methods

	T_0	T_1	T_2	....	T_k
A1	D_{10}^2	D_{11}^2	D_{12}^2	....	D_{1k}^2
A2	D_{20}^2	D_{21}^2	D_{22}^2	....	D_{2k}^2

TABLE 5-4 Results After Conversion of Table 5-3

Time	T_0	T_1	T_2	....	T_k
A1	y_{10}	y_{11}	y_{12}	....	y_{1k}
A2	y_{20}	y_{21}	y_{22}	....	y_{2k}

Total variation, S_T

$$S_T = 0^2 + y_{11}^2 + y_{12}^2 + \dots + y_{2k}^2 \tag{5.17}$$

$(f = 2k + 2)$

Variation of proportional term, S_β

$$S_\beta = \frac{(L_1 + L_2)^2}{2r} \tag{5.18}$$

$$L_1 = T_1 y_{11} + T_2 y_{12} + \dots + T_k y_{1k} \tag{5.19}$$

$$L_2 = T_1 y_{21} + T_2 y_{22} + \dots + T_k y_{2k} \tag{5.20}$$

$$r = T_1^2 + T_2^2 + \dots + T_k^2 \tag{5.21}$$

Variation of proportional constant, $S_{A\times\beta}$

$$S_{A\times\beta} = \frac{(L_1 - L_2)^2}{2r} \tag{5.22}$$

Error variation, S_e

$$S_e = S_T - (S_\beta + S_{A\times\beta}) \tag{5.23}$$

Error variance, V_e

$$V_e = \frac{S_e}{2k} \tag{5.24}$$

SN ratio, η

$$\eta = 10\log\frac{\frac{1}{2r}(S_\beta - V_e)}{V_e} \tag{5.25}$$

Sensitivity, S and S*

$$S = \frac{L_1 + L_2}{2r} \tag{5.26}$$

$$S^* = \frac{L_1 - L_2}{2r} \tag{5.27}$$

Observing S^* to compare the two treatments is the most important task. S indicates the average trend of the treatments. When S is positive, it indicates that the trend is moving toward recovery as a whole.

Instead of using Equations (5.26) and (5.27), the following β_1 and β_2 can be estimated for forecasting:

$$\beta_1 = \frac{L_1}{r} \tag{5.28}$$

$$\beta_2 = \frac{L_2}{r} \tag{5.29}$$

When β is positive, the time for recovery can be forecast by using Equation (5.15).

Section Two

MECHANICAL INDUSTRY

CHAPTER SIX

WAFER YIELD PREDICTION

6.1 OBJECTIVE

The distribution of yield from the production line is concentrated at a high-yield area and tapers down to the lower-yield area. Production management would find it useful if the yield of individual wafers could be forecasted. The yield is determined by the variability of electrical characteristics and dust. In this study, only the variability of electrical characteristics was discussed.[1]

One product was selected for study, and a Mahalanobis Space was constructed from the wafers that had high yields. The Mahalanobis Distances of various wafers were calculated in order to study the relationship between yield and the distance.

6.2 BASE SPACE

In manufacturing, functional tests are made for all semiconductor chips in a wafer, and only non-defective wafers are assembled in the next step. The yield of a wafer is the ratio of non-defective chips compared to all of the chips in a wafer.

[1]Asada, Masashi. "Yield Prediction by Using Mahalanobis Distance," Quality Engineering Forum Symposium, 1998.

Electrical characteristics are measured before the chip function is tested. From several TEG points in a wafer, 50 characteristics are measured. In this study, 100 high-yield wafers were used to measure their electrical characteristics, and a Mahalanobis Space was constructed. Instead of using all 50 characteristics, 20 characteristics were used for the construction of the Mahalanobis Space. Calculations are shown in Chapter 2, "A Detailed Example of MTS."

6.3 RELATIONSHIP BETWEEN MAHALANOBIS DISTANCE AND YIELD

The wafers, including low yield, were collected—and the Mahalanobis Distances of 20 characteristics were calculated. Table 6-1 shows the results.

From the table, it is observed that when D^2 is larger than 2, there is a possibility of having a lower yield. When D^2 is larger than 4, the possibility of having a lower yield becomes higher. The possibility exists, however, of a lower yield when D^2 is smaller than 2 and a higher yield when D^2 is larger than 4. This situation is probably caused by noise (or some causes other than electrical characteristics).

TABLE 6-1 Relationship between Mahalanobis Distance and Yield (Number of Wafers)

		Yield		
		High	Medium	Low
D^2	-2	106	7	1
	2 ~ 4	10	2	12
	+4	1	0	16
	TOTAL	117	9	27

Although this study was limited to one particular product during a certain period of time, the effectiveness of using Mahalanobis Distance for the forecasting of yield was confirmed.

6.4 SELECTION OF CHARACTERISTICS

In order to know which characteristics are useful for forecasting, we used an L_{16} orthogonal array for assignment. After examining the 20 characteristics, we decided to use six basic characteristics all the time. The remaining 14 characteristics were assigned to the orthogonal array in the following way:

- First level: do not use
- Second level: use

This relationship is shown in Table 6-2. For example, Run No. 1 involves using six basic characteristics only.

TABLE 6-2 Layout of L_{16}

															M	M1	M2	M3
A	B	C	D	E	F	G	H	I	J	K	L	M	N	e		[1]	[2]	[3]
(1)	(1)	(1)	(1)	(1)	(1)	(1)	(1)	(1)	(1)	(1)	(1)	(1)	(1)	(1)	[1]			
(1)	(1)	(1)	(1)	(1)	(1)	(1)	(2)	(2)	(2)	(2)	(2)	(2)	(2)	(2)	[2]			
(1)	(1)	(1)	(2)	(2)	(2)	(2)	(1)	(1)	(1)	(1)	(2)	(2)	(2)	(2)	[3]			
(1)	(1)	(1)	(2)	(2)	(2)	(2)	(2)	(2)	(2)	(2)	(1)	(1)	(1)	(1)	[4]			
(1)	(2)	(2)	(1)	(1)	(2)	(2)	(1)	(1)	(2)	(2)	(1)	(1)	(2)	(2)	[5]			
(1)	(2)	(2)	(1)	(1)	(2)	(2)	(2)	(2)	(1)	(1)	(2)	(2)	(1)	(1)	[6]			
(1)	(2)	(2)	(2)	(2)	(1)	(1)	(1)	(1)	(2)	(2)	(2)	(2)	(1)	(1)	[7]			
(1)	(2)	(2)	(2)	(2)	(1)	(1)	(2)	(2)	(1)	(1)	(1)	(1)	(2)	(2)	[8]			
(2)	(1)	(2)	(1)	(2)	(1)	(2)	(1)	(2)	(1)	(2)	(1)	(2)	(1)	(2)	[9]			
(2)	(1)	(2)	(1)	(2)	(1)	(2)	(2)	(1)	(2)	(1)	(2)	(1)	(2)	(1)	[10]			
(2)	(1)	(2)	(2)	(1)	(2)	(1)	(1)	(2)	(1)	(2)	(2)	(1)	(2)	(1)	[11]			
(2)	(1)	(2)	(2)	(1)	(2)	(1)	(2)	(1)	(2)	(1)	(1)	(2)	(1)	(2)	[12]			
(2)	(2)	(1)	(1)	(2)	(2)	(1)	(1)	(2)	(2)	(1)	(1)	(2)	(2)	(1)	[13]			
(2)	(2)	(1)	(1)	(2)	(2)	(1)	(2)	(1)	(1)	(2)	(2)	(1)	(1)	(2)	[14]			
(2)	(2)	(1)	(2)	(1)	(1)	(2)	(1)	(2)	(2)	(1)	(2)	(1)	(1)	(2)	[15]			
(2)	(2)	(1)	(2)	(1)	(1)	(2)	(2)	(1)	(1)	(2)	(1)	(2)	(2)	(1)	[16]			

Carefully select one wafer, each of which represents high, medium, and low-yield, respectively. These wafers are called M_1, M_2 and M_3, respectively and are assigned to outside the orthogonal array.

Calculate the Mahalanobis Distance of each wafer by using all 20 characteristics as the true values of the signal factor M_1, M_2, and M_3.

$M_1 = 0.98$

$M_2 = 1.83$

$M_3 = 3.07$

From each run (row), calculate the Mahalanobis Distances of the three wafers by using the conditions (use or do not use) indicated in each row. These results are denoted by y_1, y_2, and y_3 and are shown in Table 6-3.

TABLE 6-3 Layout and Results

														M	0.98	1.83	3.07	Average	S/N Ratio	Sensitivity	β
A	B	C	D	E	F	G	H	I	J	K	L	M	N		[1]	[2]	[3]				
														[1]	0.91	2.04	2.77	1.907	12.09	-0.41	0.95
							U	U	U	U	U	U	U	[2]	1.06	1.87	3.57	2.167	16.94	1.00	1.12
			U	U	U	U					U	U	U	[3]	0.84	1.99	3.56	2.130	14.71	0.98	1.12
			U	U	U	U	U	U	U	U				[4]	1.05	1.56	2.36	1.657	11.73	-1.85	0.81
	U	U			U	U			U	U			U	[5]	0.96	1.59	2.41	1.653	14.83	-1.74	0.82
	U	U			U	U	U	U			U	U		[6]	0.68	1.60	3.37	1.883	9.49	0.09	1.01
	U	U	U	U					U	U	U	U		[7]	0.84	1.56	3.47	1.957	9.79	0.34	1.04
	U	U	U	U			U	U					U	[8]	0.88	1.57	2.07	1.507	9.85	-2.71	0.73
U		U		U		U		U		U		U		[9]	0.97	1.76	2.63	1.787	16.30	-1.00	0.89
U		U		U		U	U		U		U		U	[10]	0.88	1.84	3.40	2.040	16.13	0.56	1.07
U		U	U		U			U		U	U		U	[11]	0.97	1.81	3.33	2.037	19.16	0.46	1.05
U		U	U		U		U		U			U		[12]	0.71	1.75	2.24	1.567	9.71	-2.14	0.78
U	U			U	U			U	U			U	U	[13]	0.95	1.96	2.26	1.723	6.65	-1.64	0.83
U	U			U	U		U			U	U			[14]	0.98	1.79	3.50	2.090	15.10	0.75	1.09
U	U		U			U		U	U		U			[15]	0.85	1.78	3.54	2.057	12.55	0.72	1.09
U	U		U			U	U			U		U	U	[16]	1.05	2.00	2.46	1.837	8.10	-1.05	0.89

U = USED BLANK = DID NOT USE

SN ratios are used to indicate the capability of forecasting. The SN ratio of No. 1 is calculated as follows.

$$S_T = y_1^2 + y_2^2 + y_3^2$$

$$= 0.91^2 + 2.04^2 + 2.77^2$$

$$= 12.6626 \qquad (f = 3) \qquad (6.1)$$

$$r = M_1^2 + M_2^2 + M_3^2$$

$$= 0.98^2 + 1.83^2 + 3.07^2 = 13.7342 \qquad (6.2)$$

$$\beta = \frac{M_1y_1 + M_2y_2 + M_3y_3}{r}$$

$$= \frac{0.98 \times 0.91 + 1.83 \times 2.04 + 3.07 \times 2.77}{13.7342}$$

$$= 0.9559 \qquad (6.3)$$

$$S_\beta = \frac{(M_1y_1 + M_2y_2 + M_3y_3)^2}{r}$$

$$= \frac{(0.98 \times 0.91 + 1.83 \times 2.04 + 3.07 \times 2.77)^2}{13.7342}$$

$$= 12.5503 \qquad (f = 1) \qquad (6.4)$$

$$S_e = S_T - S_\beta$$

$$= 12.6626 - 12.5503$$

$$= 0.1123 \qquad (f = 2) \qquad (6.5)$$

$$V_e = S_e/2 = 0.1123/2$$

$$= 0.05615 \qquad (6.6)$$

$$\eta = 10\log\frac{(S_\beta - V_e)/r}{V_e}$$

$$= 10\log\frac{(12.5503 - 0.05615)/13.7342}{0.5615}$$

$$= 12.09\ (\text{db}) \tag{6.7}$$

$$S = 10\log(S_\beta - V_e)/r$$

$$= 10\log(12.5503 - 0.05615)/13.7342$$

$$= -0.41\ (\text{db}) \tag{6.8}$$

The SN ratios of other runs are similarly calculated, as shown in Table 6-3. Figure 6-1 shows the responses of SN ratio and sensitivity.

From the responses of the SN ratio, we know that characteristics K and L are significant. From the responses of sensitivity, L is also significant. K and L belong to the same parameter; the former is average, and the latter is range.

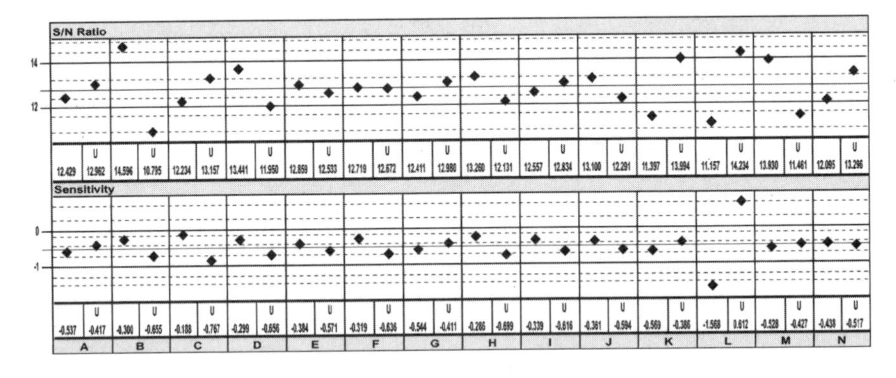

FIGURE 6-1 Response graph of SN ratio and sensitivity

Table 6-4 shows the comparison between Run No. 1 (using six basic characteristics) and the optimum condition (adding characteristics K and L).

TABLE 6-4 Results of Confirmatory Experiments

		No. 1	Optimum
Estimation	S/N Ratio Sensitivity	12.09 -0.41	17.75 1.77
Confirmation (1)	S/N Ratio Sensitivity	7.83 -1.11	11.56 1.36
Confirmation (2)	S/N Ratio Sensitivity	8.37 0.09	12.17 1.87

CHAPTER SEVEN

INKJET QUALITY INSPECTION

7.1 INTRODUCTION

Color thermal inkjet printing is an important arsenal of printing technologies for the coming years. Efficient conversion of electrical and thermal energy into robust ink drop generation, (and subsequent delivery) to a wide variety of substrates is required. Quite complex manufacturing processes are used to create devices that can deliver ink drops with required target directionality, drop velocity, drop volume, image darkness, etc. over a wide range of signal and noise space. Inspection of image quality is used during on-line production operation to cost-effectively reduce outgoing variation among units. 100 percent inspection is justified based on loss function estimates considering inspection cost, defective loss, and quality loss after shipment. Classification of print cartridges as either acceptable or not acceptable is required, based on a multitude of inspection characteristics.[1] The MTS was applied to the final inspection of image quality characteristics for a number of reasons:

(1) To summarize multivariate results with a continuous number expressing distance from centroid of "good" ones

(2) To reduce measurement costs through identification and removal of measurements with low discrimination power

[1]Lavallee, Louis. "Application of Mahalanobis Taguchi System to Thermal Inkjet Image Quality Inspection," ASI Symposium, 1998.

(3) Balance misclassification cost for both type I and type II errors

(4) Develop cast study from which practitioners and subject matter experts in training could learn the MTS method

Classification of thermal inkjet printheads as either acceptable (for shipment to customers) or rejected (for scrap) is a decision with two possible errors. A type I error occurs when a printhead is declared nonconformant when it is actually acceptable. A type II error occurs when a printhead is declared conformant when it is actually not. A type I error means that scrap costs might be high, or orders cannot be filled due to a low manufacturing process yield. A type II error means that quality loss after shipment might be high when customers discover they cannot obtain promised print quality, even with printhead priming and other driver countermeasures. Type II errors, over time, might be substantially more costly to the company than type I errors. Consequently, engineers are forced to consider a balance between shipping bad printheads and scrapping good printheads.

A Mahalanobis space was constructed by using multivariate image-quality data from 144 "good" print quality printheads. The data were used to create a Mahalanobis Space. Nonessential serial numbers were used from several months of production. In addition, the Mahalanobis Distance, D^2, of 45 nonconformant printheads was calculated by using the Mahalanobis Space. The viability of the technique to better discriminate good from bad printheads was demonstrated. In addition, from these results, the contribution rate of each measurement characteristic to the Mahalanobis Distance was calculated by using L_{64} and an analysis of variance methods. The minimum set of measurement characteristics was then selected to reduce measurement costs and to speed up the analysis process. The accelerating pace of printhead fabrication, pulled by customer demand, has placed considerable stress on the manufacturing system (which worked quite well for lower quantities of printheads). Reduction of these inspection data to one continuous Mahalanobis number and reducing measures that have low discrimination power will speed up the inspection and dispositioning process.

7.2 CAMERA INSPECTION SYSTEM

Assembled printheads were print-tested by using 100 percent inspection. During the test, power was applied to the device, ink nozzles were primed, and a variety of patterns of drops were ejected from the linear array of nozzles. A standard Xerox paper was the media that received the various patterns of drops.

A three-color line scan camera automatically captured the images of the print pattern and sent the image data in order to place templates and gather statistics on the printed patterns. A host PC initiated inspection and configuration processes, created reports, and provided a user interface for the operators.

Certain camera calibration steps were taken before testing, including white balancing (to compensate for uneven light distribution over the image plane and the spatial non-uniformity of the sensitivity of the camera's sensors). Other calibrations included printed pattern measurement and alignment (to center the image in the camera's field of view) and color dot size calibration (so that dots of differing colors are considered the same size).

Dot-aspect ratio, the ratio of a dot's height to its width, was one of the many characteristics captured by the camera system. This concept is shown in Figure 7-1. A maximum aspect ratio is allowed to control certain image-quality problems.

A second characteristic captured by the camera system was dot diameter. The diameter is an equivalent diameter calculated for noncircular dots. For each color, the mean and standard deviation of both repeated dots and same color jets were calculated. Upper and lower tolerances were applied to the dot size. Missing dot counts (or missing jet counts) were tabulated. Dot mislocation (directionality) results were summarized as both x

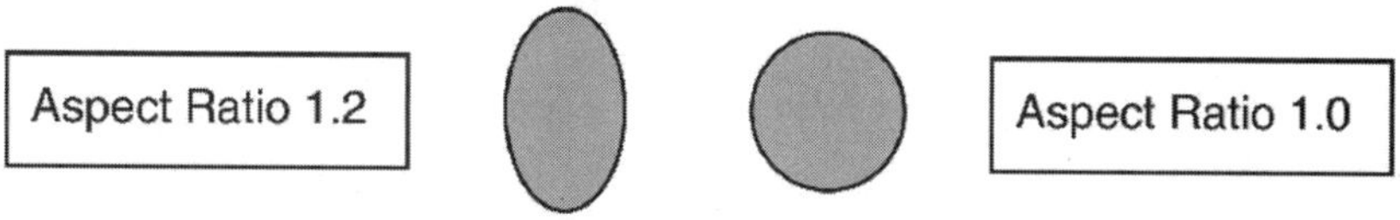

FIGURE 7-1 Dot-aspect ratio

and y error variances from least squares fitting routines of multiple dot patterns. If any single criteria limit is violated, the printhead is scrapped. Inspection results are generally reported in terms of pass/fail percentages, considering decomposition by dot, by jet, and by color for each printhead. This existing inspection process does not consider any correlation among response characteristics in decision making. Figure 7-2 shows a hypothetical plot of aspect ratio versus dot diameter. Clearly, the point labeled as abnormal (because it is outside the ellipse) would be considered normal if each characteristic were considered in isolation. This point is at $+1.4\sigma$ from the mean for spot diameter and -0.7σ from the mean of the aspect ratio. Because of correlation between the measured characteristics, one can say that this printhead has too large a spot for its aspect ratio; therefore, it should be classified as abnormal.

- With a correlation, the printhead is classified as abnormal.
- With no correlation, the printhead is classified as normal.

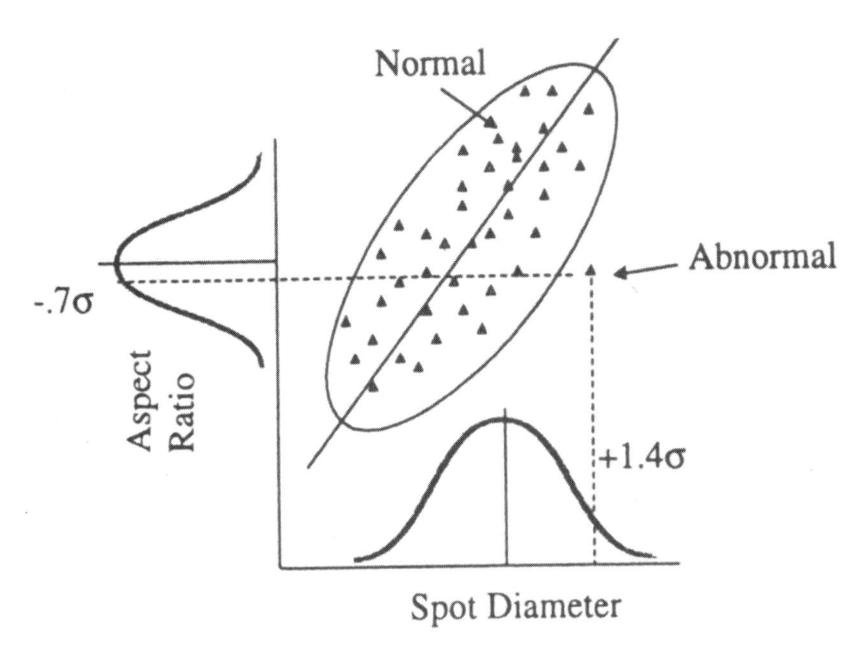

FIGURE 7-2 Correlation between spot diameter and aspect ratio

7.3 MAHALANOBIS DISTANCE RESULTS

The Mahalanobis Distance was calculated for each good printhead by using the inverse correlation matrix and values for D^2 (ranging from 0.5 to 2.5). Singular matrix problems on inversion indicated high correlations between several measured characteristics, and the elimination of one of them was required.

The average D^2 was 1.0000 for the good printheads, as expected. From these results, an origin and unit amount were established. The inverse correlation matrix from the good printheads, in turn, was used to calculate the Mahalanobis Distance for each of the 44 bad printheads. We expected that the D^2 values would be considerably larger than the D^2 values for the good printheads. These numbers ranged from 2.6 to 179.4. As a rule of thumb, a value below 4.0 for the bad printheads usually indicated a misclassification with a type I error. Those printheads with D^2 values greater than 3.0 to 4.0 can be classified as not belonging to the good group of printheads with fairly high confidence.

The MTS method enables us to distinguish normal and abnormal printhead conditions as long as we have data for the normal condition. Only the normal condition produces a uniform state, and infinite abnormal data exist. Collecting data for all those abnormal conditions is practically impossible. This method considers the degree to which the normal condition varies and produces the Mahalanobis Space. For new data, we calculate the Mahalanobis Distance, and by judging whether this distance exceeds the normal range, we can distinguish between normal and abnormal printhead conditions.

The traditional way is to discriminate to which multiple preexisting categories the actual device belongs. For printheads, one has to decide whether the printhead is good or not, a large volume of defective condition data and normal condition data are measured beforehand, and a database is constructed with both. For newly measured data (for which we do not know whether the printhead is good or bad), we calculate the group to which the data belongs.

In reality, however, there are a multitude of image-quality problems. So if information about various types of images is not obtained beforehand, printheads cannot be distinguished. Because the traditional discrimination method requires this information, it is not practical.

Table 7-1 partially shows a correlation matrix of the 58 observed characteristics. Table 7-2 shows the calculated D^2 values of the 145 good printheads. Table 7-3 shows the Mahalanobis Distance for the 44 bad printheads, using the inverse correlation matrix from the good printheads.

Figure 7-3 shows the Mahalanobis Distances D^2 for both the good and bad printheads. A wide range of D^2 values were observed for the non-conforming printheads. Note that the good printheads are quite uniform, which is usually the case. Only one

TABLE 7-1 Correlation Matrix (Part)

	1.00	-0.24	0.24	-0.25	0.12	-0.15	-0.03	-0.24	0.34	0.24	0.37	0.44	0.42
:	-0.20	-0.10	0.18	0.18	-0.29	0.30	-0.27	0.05	0.07	0.03	-0.08	0.29	0.38
:	0.25	-0.05	-0.15	0.21	0.45	-0.17	-0.05	-0.01	0.01	-0.00	0.02	0.27	0.24
:	0.37	0.25	0.04	-0.13	0.24	0.03	-0.19	0.07	-0.18	-0.03	0.04	-0.02	0.06
:	-0.04	0.20	0.16	0.06	-0.16	0.27							
	-0.24	1.00	0.08	0.03	0.19	0.13	0.25	-0.18	-0.42	0.18	0.12	-0.40	-0.32
:	0.03	-0.24	-0.15	-0.31	0.97	0.06	0.14	-0.26	0.16	-0.25	0.11	0.02	-0.41
:	-0.28	0.05	-0.24	-0.21	-0.11	0.94	0.01	0.06	0.02	0.06	0.01	-0.06	0.07
:	-0.35	-0.29	0.27	-0.28	-0.20	-0.29	0.83	-0.15	0.14	-0.19	0.23	-0.31	0.07
:	0.08	-0.37	-0.28	-0.02	-0.25	-0.22							
	0.24	0.08	1.00	-0.53	0.44	-0.21	-0.11	-0.35	-0.02	0.21	0.22	0.02	-0.01
:	0.16	-0.10	-0.06	-0.18	0.08	0.29	-0.35	-0.09	0.07	-0.15	-0.30	0.05	-0.03
:	-0.05	0.14	-0.16	-0.05	0.13	0.15	0.08	0.08	-0.22	0.08	-0.12	0.26	0.09
:	-0.07	-0.11	0.29	-0.15	-0.02	-0.17	0.04	0.21	-0.32	-0.18	0.28	-0.18	0.21
:	0.06	-0.18	-0.07	-0.03	-0.19	-0.02							
	-0.25	0.03	-0.53	1.00	-0.20	-0.02	-0.23	0.29	-0.10	-0.34	-0.33	-0.26	-0.16
:	0.04	0.24	-0.09	-0.03	0.05	-0.28	0.82	0.00	-0.09	-0.13	0.14	-0.11	-0.20
:	-0.09	-0.01	0.28	-0.10	-0.13	0.00	-0.15	0.29	0.37	-0.17	-0.04	-0.34	-0.05
:	-0.11	-0.02	-0.10	0.24	-0.12	0.03	0.11	-0.28	0.72	0.04	-0.23	0.06	-0.27
:	-0.01	0.02	0.01	0.09	0.31	-0.12							
	0.12	0.19	0.44	-0.20	1.00	-0.26	-0.05	-0.08	-0.21	0.31	0.17	-0.13	0.02
:	-0.01	-0.06	-0.09	-0.14	0.15	0.28	-0.08	-0.05	-0.05	-0.15	-0.05	0.11	-0.22
:	-0.16	0.05	-0.07	-0.12	0.13	0.25	-0.06	0.13	0.03	-0.13	0.08	0.17	0.11
:	-0.08	-0.01	0.19	-0.06	-0.12	-0.10	0.22	-0.07	-0.06	-0.12	0.07	-0.08	0.12
:	0.05	-0.12	-0.10	0.07	-0.08	-0.12							
	-0.15	0.13	-0.21	-0.02	-0.26	1.00	0.48	0.20	-0.16	-0.07	0.01	-0.25	-0.23
:	-0.04	0.07	-0.02	-0.01	0.11	0.04	-0.08	0.03	0.67	0.38	0.19	0.10	-0.18
:	-0.09	0.06	0.11	-0.05	-0.25	0.10	0.04	-0.50	-0.02	0.53	0.14	0.01	0.03
:	-0.20	-0.08	-0.09	0.11	-0.07	0.10	0.01	0.12	-0.28	0.12	0.14	0.12	-0.05
:	-0.08	-0.10	-.01	0.01	0.12	-0.08							
	-0.03	0.25	-0.11	-0.23	-0.05	0.48	1.00	0.11	-0.10	0.31	0.26	-0.11	-0.10
:	-0.12	-0.10	-0.01	-0.11	0.22	0.08	-0.11	-0.12	0.14	0.17	0.28	0.02	-0.11
:	-0.04	0.03	-0.06	-0.06	-0.09	0.23	0.00	-0.14	-0.14	0.10	0.25	0.07	0.03
:	-0.07	-0.04	0.05	-0.07	-0.04	-0.05	0.28	-0.18	-0.15	-0.04	-0.04	0.07	0.13
:	-0.01	0.02	0.02	0.03	-0.07	-0.05							

TABLE 7-2 Mahalanobis Distance from Good Printheads

	0.9788539	0.7648473	0.8365425	1.1541308	1.5237813	0.6360767	1.9261662
:	0.6253195	1.8759381	0.6151459	0.8409623	2.3004483	0.8254882	0.9997392
:	0.8141697	1.0554864	1.3502383	1.4477616	1.575254	0.6408639	0.9636183
:	2.0532355	1.2753139	1.0992817	0.9767287	0.7537741	0.7801704	0.6107221
:	0.6377904	1.1194686	0.7462556	1.0316818	1.8069662	1.9241475	1.0319332
:	0.9899644	1.2182822	1.2182822	1.1331171	0.8370348	2.2826932	0.9207974
:	1.1467803	1.0811343	1.005008	0.9170464	1.2295005	0.5892113	0.4927686
:	0.8277777	0.9806735	0.9319074	0.9434869	0.8671445	0.8750086	0.8219719
:	1.2870564	0.6186984	1.5087154	1.0080104	1.343378	1.3982634	1.338509
:	0.5055908	0.9222672	1.2616672	2.0557088	0.4774843	0.7116405	0.4109008
:	0.9073186	1.0098421	0.610628	1.8020641	0.7664054	0.5304222	1.22455
:	0.6284057	1.407896	0.9961997	0.6662852	0.8032502	1.2031408	0.9997437
:	1.2198565	1.0614666	0.5896821	0.4663961	1.63111	0.7763159	1.0746786
:	0.744278	0.5047632	1.0467326	1.0163124	0.9345957	0.6408224	0.5880229
:	1.3450012	0.7046379	1.4418307	0.7789877	1.0798978	0.6764585	0.8649927
:	0.4805313	0.4805313	1.0123116	0.7207266	1.3288415	0.6955458	0.9248074
:	1.2550094	0.9059842	1.1129645	0.7871329	0.6158813	0.762891	0.7765336
:	0.5245411	0.5759779	1.0465223	1.2335352	1.6694431	0.9200819	1.4373301
:	0.8082051	0.9922023	0.9889715	0.5822106	1.5744132	0.5348402	1.0790753
:	0.6256058	0.8307524	0.8303376	1.3077433	0.6317183	1.2935359	0.509095
:	1.2105074	1.1197155	0.7580503	0.9266635	0.6304593		

TABLE 7-3 Mahalanobis Distance from Bad Printheads

	22.267015	6.3039776	47.516228	86.911572	15.757659	47.675354	9.2641579
:	2.5930287	14.052908	11.691446	15.090696	13.894336	8.6390361	179.43874
:	20.041785	10.695167	29.33736	13.926274	34.2429	9.3282624	44.442126
:	35.30109	18.388437	55.787193	6.0059889	63.12819	16.813476	45.355212
:	78.914301	29.200082	42.105292	28.056818	46.802471	121.26875	8.1542113
:	33.730109	9.3395192	6.77362	22.493329	16.492573	4.6645559	26.666635
:	22.46681	153.76889					

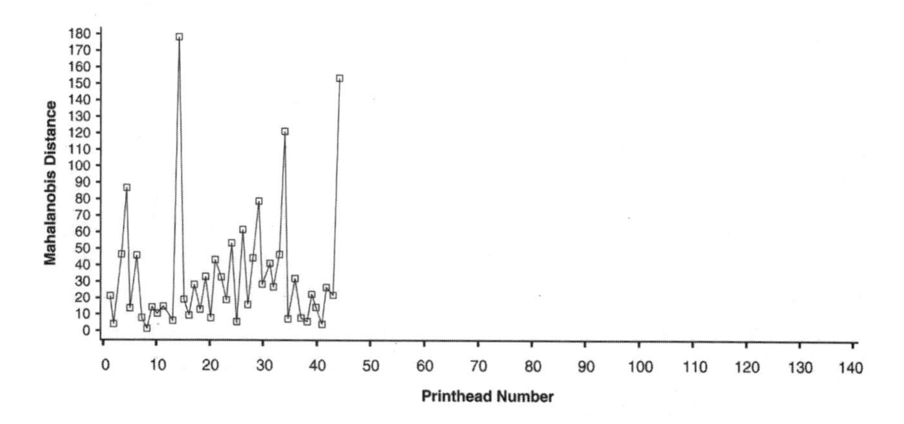

FIGURE 7-3 Mahalanobis Distance versus printhead number for both passing and failing printheads

or two cases of printheads is classified as bad. The threshold value for D^2 is determined by economic consideration. Misclassification countermeasure costs are usually considered.

7.4 MEASUREMENT SYSTEM COST REDUCTION

Among the 58 different measured characteristics in this example, there were a number that were doing most of the discrimination work. An L_{64} orthogonal array was used to identify those characteristics. In this case, the orthogonal array is used as a decision tool. For each column of the L_{64} orthogonal array, one measured characteristic was assigned to each column, using 58 of the 63 available columns. L_{64} is a two-level design with 63 degrees of freedom for assignment. A value of 1 (or Level 1) in any cell of the standard array layout indicated that the measured characteristic was considered, while a value of 2 (or Level 2) meant that the measured characteristic was not considered. In Row 11, for example, 27 of the 58 characteristics that were measured were considered, because there were 27 Level 1 characteristics. Using those 27 columns, the analysis was conducted in order to generate a Mahalanobis Space with the good printheads, and Mahalanobis Distances for each of the 44 bad printheads were calculated. From these 27 characteristics and 44 bad printheads, a single SN ratio (the larger-the-better characteristic) was calculated according to the following equation:

$$\eta = -10\log\frac{1}{m}\left(\frac{1}{D_1^2}+\frac{1}{D_2^2}+\ldots+\frac{1}{D_m^2}\right) \qquad (7.1)$$

where summation is from m = 1 to m = 44. This same procedure is followed for the remaining L_{64} rows. A larger-the-better form was used, because the intent was to amplify the discrimination power by selecting a subset of measures. This action, in turn, would reduce the cost and time necessary to process inspection data. The results are shown in Table 7-4 for each of the L_{64} rows. The SN ratio ranged from −16.6 db to −4.7 db. Table 7-5 shows the Mahalanobis Distances for each row of L_{64}.

TABLE 7-4 SN Ratios for Each L_{64} Combination

EXPT	S/N RATIO
1	-16.1455
2	-15.1823
3	-13.4513
4	-14.7418
5	-13.0265
6	-13.9980
7	-14.8928
8	-15.1808
9	-14.3932
10	-15.7381
11	-16.0573
12	-13.2949
13	-11.3572
14	-12.5372
15	-14.0662
16	-15.9616
17	-10.4232
18	-12.4296
19	-14.1015
20	-11.5102
21	-15.1508
22	-14.3844
23	-12.2187
24	-15.3981
25	-13.3026
26	-9.4273
27	-12.2470
28	-15.1932
29	-14.4695
30	-14.4875
31	-14.8804
32	-14.6857
33	-13.7804
34	-15.8187
35	-11.5528
36	-14.3503
37	-14.3373
38	-14.3891
39	-14.0671
40	-14.8759
41	-15.3622
42	-15.4632
43	-12.5654
44	-13.9813
45	-16.3452
46	-13.0218
47	-15.3362
48	-15.8292
49	-12.2162
50	-14.5625
51	-15.3548
52	-16.6430
53	-13.9844
54	-8.9751
55	8.3045
56	-13.8523
57	-11.0553
58	-14.9251
59	-15.4196
60	-11.7635
61	-9.8751
62	-14.5514
63	-13.5409
64	-4.7039

TABLE 7-5 Mahalanobis Distance for Each Row of L_{64}

Row 1 of L_{64}	3.3948335	0.7672923	3.7212996	7.5160489	2.1628283	0.4735541	0.9902953
:	0.8921714	0.8890575	0.4336029	0.7916304	3.1689748	0.3589341	15.351063
:	1.6113871	0.8156152	5.1049158	3.53441	0.4765632	0.6596119	0.5178774
:	7.8548821	0.6789542	2.3212435	1.1780486	0.620615	4.3913711	30.419276
:	1.3079055	1.5259677	0.5979884	0.9019846	4.0599383	6.7014764	0.35931
:	0.6227532	3.3683819	1.3396863	5.3121466	0.7085592	0.7498931	11.770036
:	1.7869636	2.6817723					
Row 2 of L_{64}	2.1538859	0.5691297	5.0493192	6.0833953	2.26442	1.1433147	1.1618087
:	0.8776536	0.743007	0.861313	0.5233098	2.0991015	1.2277685	14.323771
:	3.5200479	0.8022022	4.8885268	2.330918	0.8417254	1.1887293	0.7514819
:	4.1954716	6.0096008	1.0382971	3.080755	1.5613432	2.0078949	3.9968699
:	0.9804545	2.4407172	0.5850997	1.1164188	1.5777033	7.4701672	0.5262845
:	0.7533587	1.1707115	1.3342356	2.7278281	1.0099527	0.8908078	12.579455
:	2.772986	7.0534854					
Row 3 of L_{64}	2.432585	1.8731996	6.2922649	8.802438	2.2404416	1.631539	1.9070269
:	1.0529453	1.2410093	1.0671635	4.211971	2.7976614	0.7359613	6.9516176
:	1.2869917	3.3668795	4.4021798	5.6621817	1.9823413	2.2946022	0.7033499
:	14.94987	1.6268848	2.0984149	1.7826083	1.5005839	7.4753245	25.570186
:	0.9018103	7.125921	2.3458211	1.0736915	6.0013726	31.945727	3.6939773
:	3.2325816	2.7765327	0.5454912	3.371364	1.5793315	1.2316633	10.05149
:	2.5569832	2.9596748					
↓							
Row 64 of L_{64}	22.267015	6.3039776	47.516228	86.911572	15.757659	47.675354	9.2641579
:	2.5930287	14.052908	11.691446	15.090696	13.894336	8.6390361	179.43879
:	20.041785	10.695167	29.33736	13.926274	34.2429	9.3282624	44.442126
:	35.30109	18.388437	55.787193	6.0059889	63.12819	16.813476	45.355212
:	78.914301	29.200082	42.105292	28.056818	46.802471	121.26875	8.1542113
:	33.730109	9.3395192	6.777362	22.493329	16.492573	4.6645559	26.666635
:	22.46681	153.76889					

An analysis of variance procedure was conducted on these 64 SN ratios in order to break the total variation into constituent parts. The results showed a handful of measures; e.g., C_2, C_8, C_{15}, C_{17}, C_{18}, C_{19}, C_{29}, C_{43}, and C_{57} have larger effects, and the results also showed the small effects of many others. This conclusion suggests that not all of these measures are adding value to the decision to scrap or ship printheads. Much of the time, the measurements add little or nothing to the decision. Substantially reducing the number of measures reduces the time and cost of inspection and dispositioning.

7.5 CONCLUSIONS

MTS was applied to the analysis of image-quality data collected during final inspection of thermal inkjet printheads. A

Mahalanobis Space was constructed by using results from a variety of printheads classified as "good" by all criteria, establishing an origin and unit number. Mahalanobis Distance for non-conforming printheads was calculated and demonstrated the capability to discriminate between normal and abnormal printheads. Among the 58 product image-quality characteristics that were investigated, only a fraction of the total were doing the work of discriminating good from bad for actual production, identifying the opportunity to reduce measurement and overhead cost for the factories.

Misclassification of printheads was checked, and as expected, the scrapping of "good" printheads was more prevalent that shipping "bad" printheads. The scrap cost was approximately one-tenth the cost for a type-II error, so naturally, one would be more likely to scrap good printheads. The opportunity to apply MTS is being explored for asset recovery. For example, when copiers and printers are returned from customer accounts, recovery of motors, solenoids, clutches, lamps, etc. (which are still classified as "good") can be reused in new or refurbished machines.

CHAPTER EIGHT

PREVENTION OF DRIVING ACCIDENTS

8.1 INTRODUCTION

In the automotive industry, researchers are working to develop a sensing system to help prevent driving accidents. Such a sensing system would treat time-series, multi-dimensional information (such as changing the car's speed or distance); therefore, MTS would come in handy. Researchers conducted a basic study in order to determine whether it is possible to judge dangerous automobile situations and to determine what kind of judging ability can be expected.[1]

8.2 MEASURING SYSTEM

Sensors that are necessary for an accident-prevention system are classified into two groups: the sensors that recognize the environment around a car, and the sensors that detect the conditions of a car. Because this study is basic, the most basic sensors (distance sensors and a speed sensor) were used. Figure 8-1 shows the measuring system.

One laser radar was used in the front, and ultrasonic sensors were used at the front right, front left, and rear positions in order to measure distances. For the speed, we used a speed sensor. The maximum distance measurable was 150 meters for the laser radar and

[1]Mizoguchi, Kazutaka, Akiko Kawai, Kazutaka Hamada (Nissan Motor Co.), Takashi Kamoshita (National Research Laboratory of Metrology). "A Research on a Sensing System Using Mahalanobis Distance for Prevention of Driving Accidents," Quality Engineering Forum Symposium, 1998.

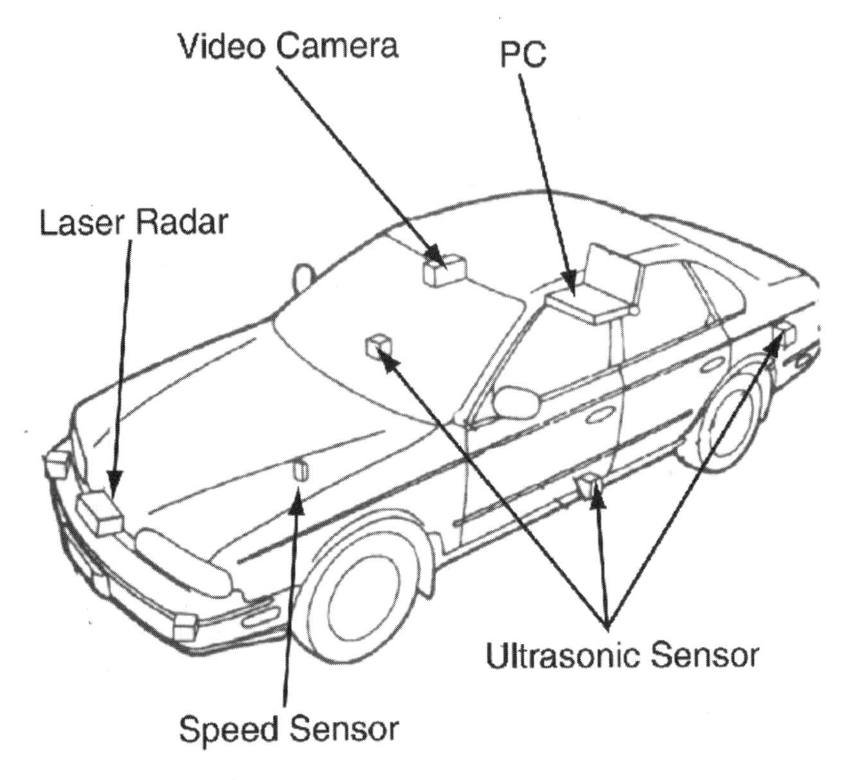

FIGURE 8-1 Measuring system

six meters for ultrasonic sensors. The laser radar scans the objects within a 12-degree angle by emitting 80 lasers in 150 milliseconds. An ultrasonic sensor measures the distance of the closest object within a 10-degree angle. When the radar did not sense any objects, researchers used 150 meters for the laser radar and six meters for ultrasonic sensors. With a measuring cycle of 150 milliseconds, 33 time-series pieces of data were recorded in a 4.8-second period.

8.3 BASE SPACE

The definition of a base space depends on what kind of action will follow the results. Determining whether a specific condition is dangerous depends on the application. Examples of applications are as follows:

(1) giving an alarm to the driver so that the driver can take precautionary measures
(2) using an automatic precaution system, such as using the brake or the steering wheel

In this study, application type (2) was considered as a base. A safe driving situation was defined as a situation in which nothing was coming close to the car.

8.4 BASE DATA COLLECTION

Data were collected when the car was running on public roads under different road shapes, a different number of lanes, or at a different speed.

If data were collected from seven sensors with 33 time-series data, the total number of characteristics would be 2838[(80 + 6) × 33], which is too many. In order to reduce the number of characteristics, 80 laser beams were divided into 20 (four in a group), and researchers recorded the minimum value of each group. In this way, the number of characteristics was reduced from 86 to 26. For the time-series data, seven points (4.8 seconds before, 2.4 seconds before, 1.2 seconds before, 0.6 seconds before, 0.3 seconds before, 0.15 seconds before, and the current time) were used.

In the calculation of Mahalanobis Distance as described in the following section, the correlation coefficient of car speed and the rear ultrasonic sensor showed values as high as 0.99, and its element in the inverse correlation matrix had a number of more than four digits (refer to Table 8-1). For this reason, only the current time for car speed was used. Also, the values of current time for car speed and 4.8 seconds before for the rear ultrasonic sensor were used, suggesting that time-series data of car speed and the rear ultrasonic sensor should not be used together.

The researchers decided to construct a base space by using 117 characteristics, including the sensors and time-series items (as shown in Table 8-2).

TABLE 8-1 Test-Driving Places and Number of Data Sets

	Laser Radar (1)		Laser Radar (1)	
	Mid/High Speed	Low Speed*	Mid/High Speed	Low Speed*
Straight	85	111	54	44
Curve	32	6	26	3
Dividing/Merging	20	33	3	1
Interstate	---	---	7	18
Others	0	5**	0	0
TOTAL	**137**	**155**	**90**	**66**

* Low speed is under 60 km/hr for car only road and under 20 km/hr for ordinary road.
** 3 sets measured at toll plazas and 2 sets measured at service areas.

TABLE 8-2 Examples of Data (Part)

	Laser Radar (1)								Ultrasonic Sensor	
	Before 4.8 sec.	Before 2.4 sec.	Before 1.2 sec.	Before 0.6 sec.	Before 0.35 sec.	Before 0.15 sec.	Now	....	Before 4.8 sec.	Now
No. 1	3.00	3.00	3.00	3.00	3.00	3.00	3.00	....	2.63	2.60
No.2	150	17.7	150	150	150	150	150	....	6.00	6.00
No.3	150	150	150	150	150	150	150	....	6.00	6.00
↓	↓	↓	↓	↓	↓	↓	↓	....	6.00	6.00

8.5 BASE SPACE CONSTRUCTION AND MAHALANOBIS DISTANCE DISTRIBUTION

The construction of the Mahalanobis Space and calculation of the Mahalanobis Distance are shown as follows:

(1) From the measured results X_{ij} (characteristics i = 1 ~ 171, number of data sets j = 1 ~ 448), calculate the average, m_i, and standard deviation, σ_i, of each characteristic (as shown in Table 8-3). The data in Table 8-3 are normalized by using the following equation (refer to Table 8-4):

$$x_{ij} = \frac{X_{ij} - m_i}{\sigma_i} = \frac{3 - 100}{60.3} = -1.610 \qquad (8.1)$$

(2) Calculation of the correlation matrix and inverse matrix

From x_{ij}, you can construct the correlation matrix R and its inverse matrix, R^{-1}.

TABLE 8-3 Average and Standard Deviation (Part)

	Laser Radar (1)								Ultrasonic Sensor	
	Before 4.8 sec.	Before 2.4 sec.	Before 1.2 sec.	Before 0.6 sec.	Before 0.35 sec.	Before 0.15 sec.	Now	....	Before 4.8 sec.	Now
Avg.	100	101	102	102	104	104	104	....	5.60	5.70
Std. Dev.	60.3	59.6	59.7	60.0	59.0	59.1	59.7	....	1.18	1.03

TABLE 8-4 Normalized Data (Part)

	Laser Radar (1)								Ultrasonic Sensor	
	Before 4.8 sec.	Before 2.4 sec.	Before 1.2 sec.	Before 0.6 sec.	Before 0.35 sec.	Before 0.15 sec.	Now	....	Before 4.8 sec.	Now
No. 1	-1.61	-1.65	-1.67	-1.66	-1.71	-1.70	-1.69	....	-2.50	-3.01
No.2	0.83	-1.40	0.80	0.79	0.78	0.79	0.77	....	0.3	0.29
No.3	0.83	0.82	0.79	0.79	0.78	0.79	0.77	....	0.34	0.29
↓	↓	↓	↓	↓	↓	↓	↓	....	↓	↓ ....

$$R = \begin{bmatrix} 1 & 0.36 & \cdots & 0.31 \\ 0.36 & 1 & \cdots & 0.32 \\ \cdots & \cdots & \cdots & \cdots \\ \cdots & \cdots & \cdots & \cdots \\ 0.31 & \cdots & \cdots & 1 \end{bmatrix} \quad (8.2)$$

$$R^{-1} = \begin{bmatrix} 2.91 & -0.28 & \cdots & 0.65 \\ -0.28 & 2.62 & \cdots & 0.02 \\ \cdots & \cdots & \cdots & \cdots \\ \cdots & \cdots & \cdots & \cdots \\ 0.65 & 0.02 & \cdots & 7.36 \end{bmatrix} \quad (8.3)$$

(3) Calculation of Mahalanobis Distance

From the elements of the inverse matrix, $a_{ii'}$ ($i, i' = 1 \sim 171$), the Mahalanobis Distance, D_j^2 ($j = 1 \sim 448$), is calculated by using the following equation:

$$D_j^2 = \frac{1}{171} \Sigma \Sigma \, a_{ii'} \, x_{ij} \, x_{i'j} \quad (8.4)$$

Figure 8-2 shows the distribution of Mahalanobis Distance of 448 sets of data. The distribution has an average of 1. All results are fewer than 2.

In order to determine whether the base space is appropriate, several sets of data that belong to safe driving were used to calculate the Mahalanobis Distance. All results were fewer than 2, which proved that the base space was appropriate. Figure 8-3 shows an example.

8.6 MAHALANOBIS DISTANCE UNDER DANGEROUS SITUATIONS

The Mahalanobis Distance changes of some typical situations are observed as follows:

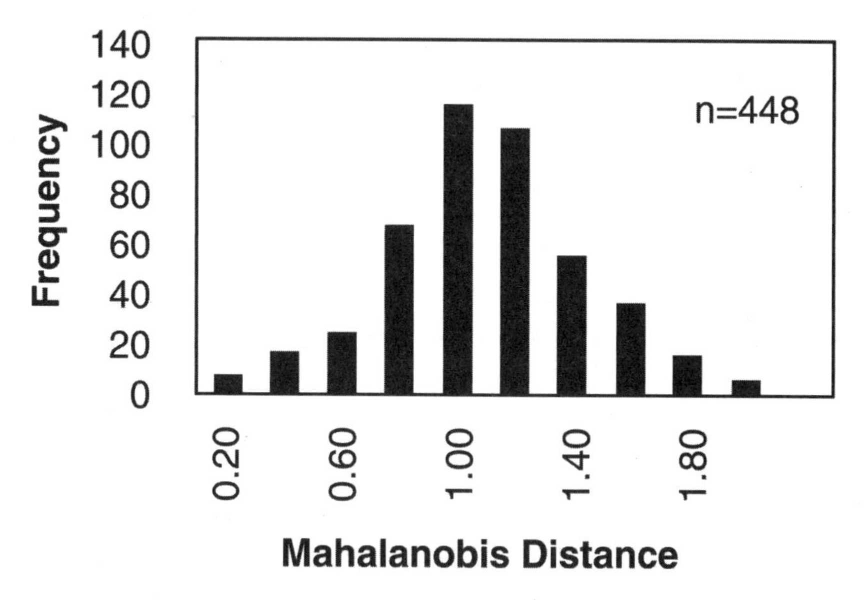

FIGURE 8-2 Mahalanobis Distance of the Base Space

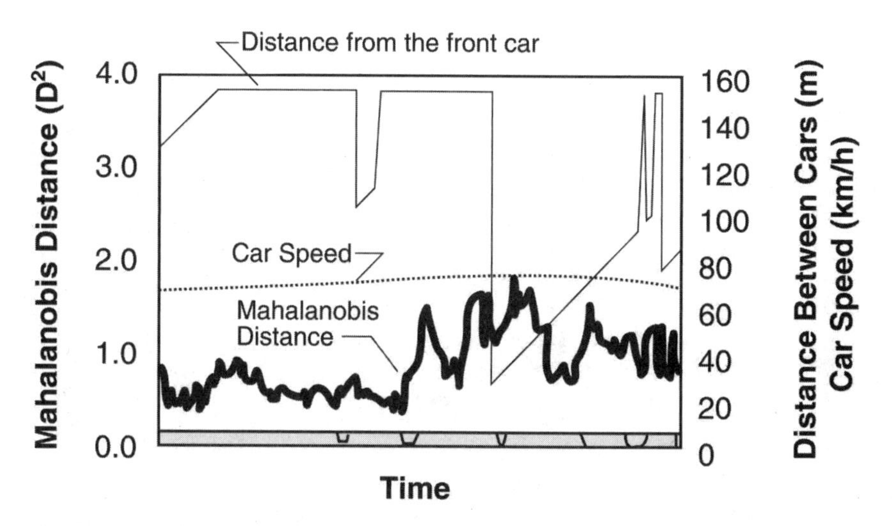

FIGURE 8-3 Base space

- (Scene 1)—When a car is getting close to the front car, its Mahalanobis Distance increases. After the car starts to reduce its speed, its Mahalanobis Distance decreases with a small period of time lag—indicating that the car is becoming safer. Therefore, the extent of danger can be expressed by Mahalanobis Distance (refer to Figure 8-4).
- (Scene 2)—Another car cut in front of the car at a relatively small speed difference. Clearly, a driver should be made aware before another car cuts in front of him or her. When the distance between two cars is short, the situation is dangerous. Mahalanobis Distance can express this danger situation (refer to Figure 8-5).
- (Scene 3)—A car collided with a still object at a speed of 30 km/hr. The Mahalanobis Distance exceeded 2 before the collision; therefore, it was possible to detect the danger before the collision. The value was smaller than the values without a collision, however. The reasons might be as follows:

  (1) The experiment was conducted in test courses, where no structures existed in the environment.

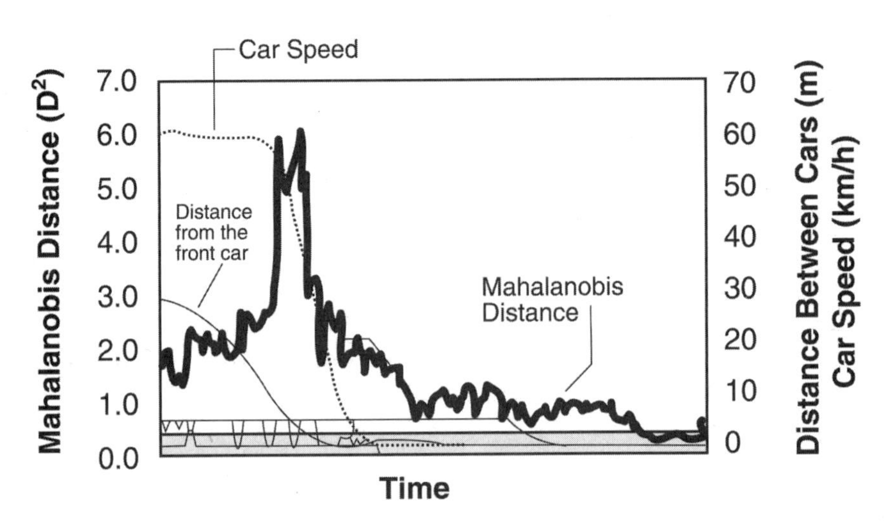

FIGURE 8-4 Mahalanobis Distance change of scene No. 1

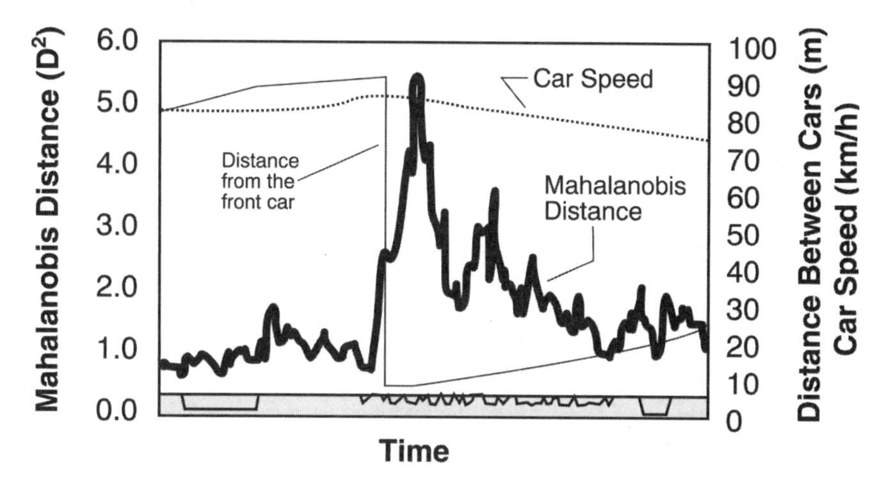

FIGURE 8-5 Mahalanobis Distance change of scene No. 2

(2) The car speed was 30 km/hr, which is slower than the average speed that was used to construct the base space.

Conducting the experiment on a public highway would be more beneficial, but this environment is impossible. The simulation trial has not yet begun (refer to Figure 8-6).

8.7 EVALUATION OF FUNCTIONALITY

Twenty-six characteristics exist, including 20 for the laser radar, five for the ultrasonic distance sensors, and one for the speed sensor. To improve the efficiency and capability of discrimination, researchers studied the effects of the characteristics. The speed sensor and five ultrasonic sensors, including front right, front left, two sides, and rear distance, were used throughout the study. The study dealt with the changes that were due to the changes in laser beams. Ten characteristics, (1–20), (2–19), ..., and (10–11), as shown in Figure 8-7, were assigned to an L_{12} orthogonal array.

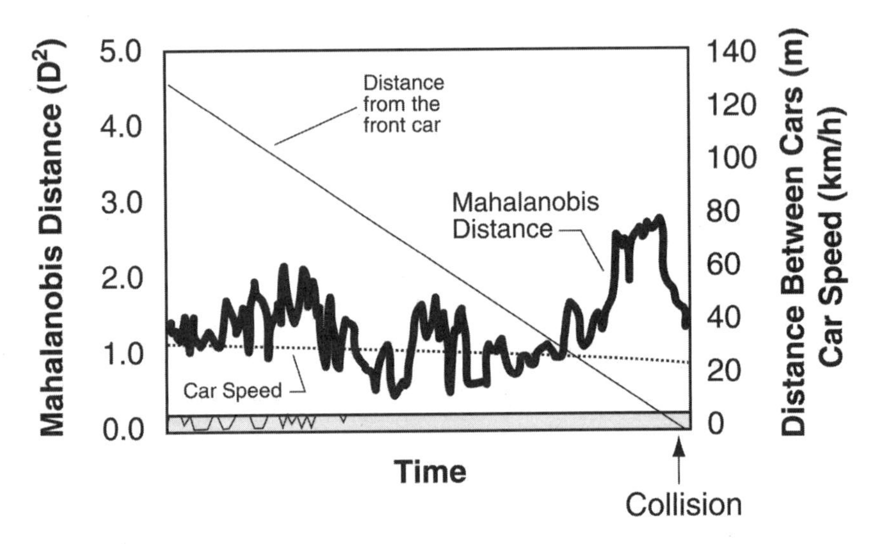

FIGURE 8-6 Mahalanobis Distance change of scene No. 3

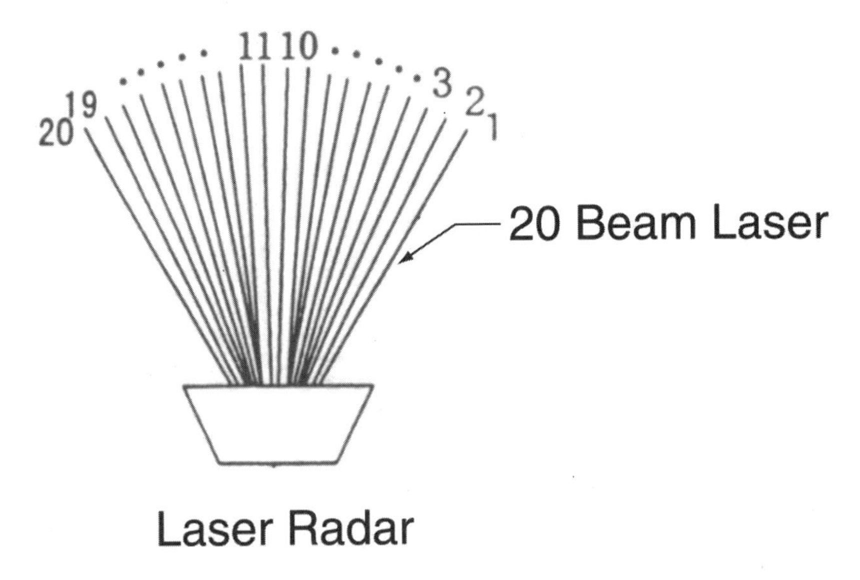

FIGURE 8-7 Laser item number

In the array, Level 1 is marked "to use," and Level 2 is marked "not to use." Therefore, Run No. 1 indicates "to use all characteristics." Sixty sets of peaks from the data, indicating

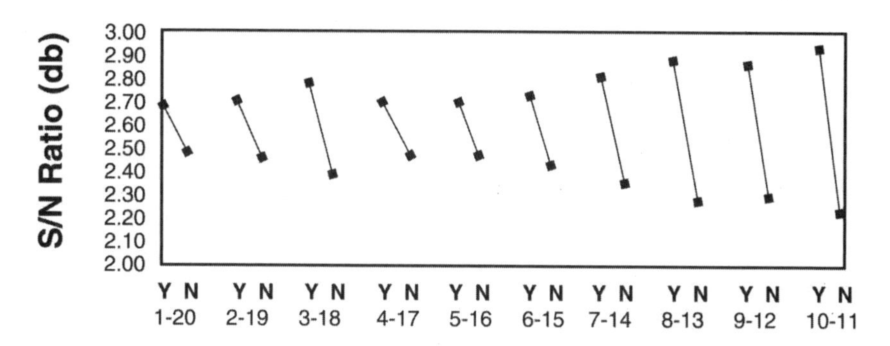

FIGURE 8-8 Factorial effects

dangerous situations as described previously, were collected. From each condition of the L_{12} array, researchers calculated SN ratios of larger-the-better characteristics. The factorial effects are shown in Figure 8-8.

In Figure 8-8, all characteristics showed that "use" is better than "do not use." The figure also indicates that the beams that are closer to the center have a higher discriminating power. Figures 8-9, 8-10, and 8-11 compare the discriminating power between "using all beams" and "without using smaller effects: (1–20), (2–19), (4–17), and (5–16)."

Researchers concluded from the comparison that in any scenes, the discriminating power decreased for the latter case. Therefore, these laser beams cannot be reduced.

8.8 CONCLUSION

In the study, the base space was defined as the situation in which no objects were getting closer to the car. Based on this space, it was possible to recognize a dangerous situation, such as another car cutting in or a collision with a still object. The study had just begun, however. The following list describes areas that are to be further studied:

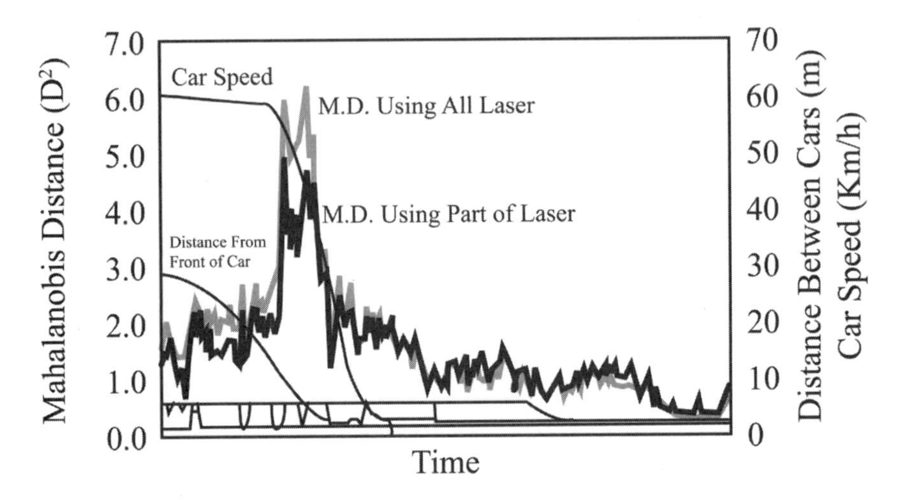

FIGURE 8-9 Mahalanobis Distance change of scene No. 1

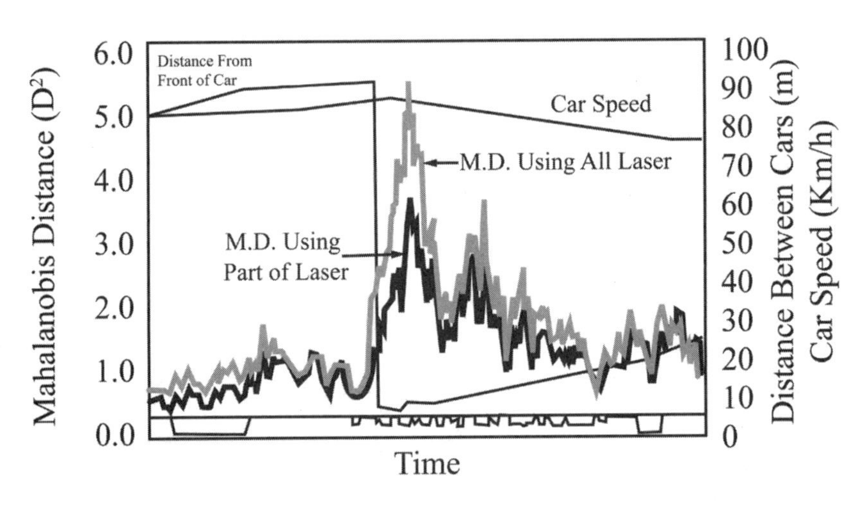

FIGURE 8-10 Mahalanobis Distance change of scene No. 2

(1) Selection of systems. Beside speed sensor and distance sensors, which were used as a basic study, other sensors exist, such as steering sensors or wheel rotating speed sensors. By adding these sensors, the recognition power could be improved.

(2) Distance sensor types. In this study, a laser radar and ultrasonic sensors were used. The sensors that use

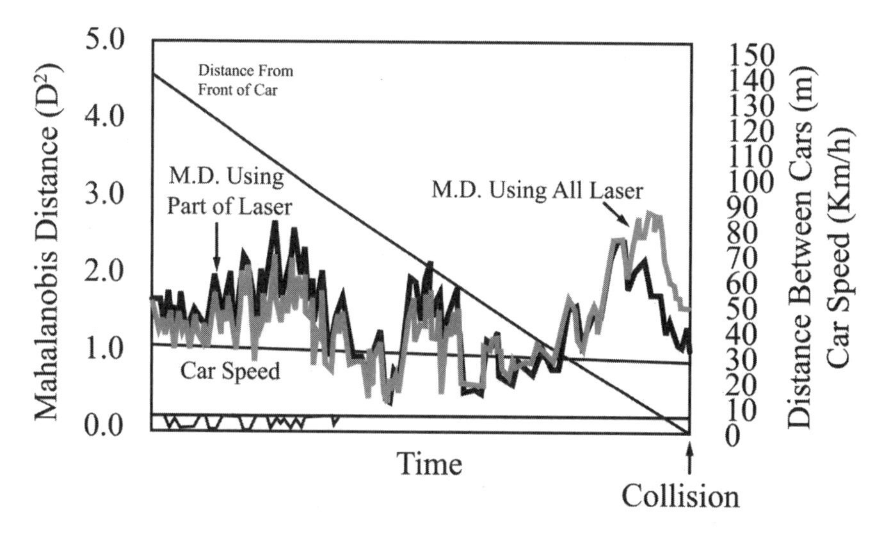

FIGURE 8-11 Mahalanobis Distance change of scene No. 3

electronic wave or picture images are currently being developed. These types could be used in the future.

(3) Data treatment. In this study, 80 laser beams were reduced to 20. The total number of characteristics was 171, however, which is not a small number for the computer that is installed inside a car.

(4) Definition of the base space. Based on the definition in this study, we can label objects such as sign boards as dangerous.

SECTION THREE

ELECTRICAL INDUSTRY

CHAPTER NINE

SOLDER JOINT APPEARANCE INSPECTION

9.1 INTRODUCTION

The automation of solder joint appearance inspection is being studied and applied in many companies. Several systems are available, and each company adopts the system that fits its particular requirement and environment.

Since 1990, Seiko Epson Corporation has started to introduce solder joint automatic inspection for the circuit boards of personal computers, printers, or liquid-crystal projectors.[1] Such devices utilize inspection logic in the calculation, based on the measured data, and the results are compared with the specified standards in order to determine whether the circuit board passes or fails. Two types of errors exist, however: good products being judged as defective, and defective products being judged as good. Efforts have been made to reduce these errors by adjusting the judging standards of the inspection logic. Such work requires a high level of professional knowledge. In this study, a product called *Quad Flat Package* (QFP) that has fine pitches and causes frequent inspection errors was used for inspection.

[1]Makabe, Akira, Kei Takada (Seiko Epson Corp.) Hiroshi Yano (Japanese Standards Association). "An Application of the Mahalanobis Distance for Solder Joint Appearance Automatic Inspection," Quality Engineering Forum Symposium, 1998.

Figure 9-1 shows the principle of laser-type appearance inspection. Solder joints are scanned by the He-Ne laser. Its reflection angle varies due to the angle of the solder joint, and a series of numbers are recorded. The following is an example of a series of numbers:

666666666655444333234466666

The numbers were recorded in the order of time. The larger the number, the flatter the surface. The steepest surface in these numbers is 2. Figure 9-2 shows the cross-section of a soldered joint. Figures 9-3 and 9-4 show the images of good and defective joints that were drawn after scanning.

9.2 DATA COLLECTION AND MAHALANOBIS DISTANCE CALCULATION

Eleven defective and 2,670 non-defective QFP leads from the manufacturing process were collected for analysis. In order to

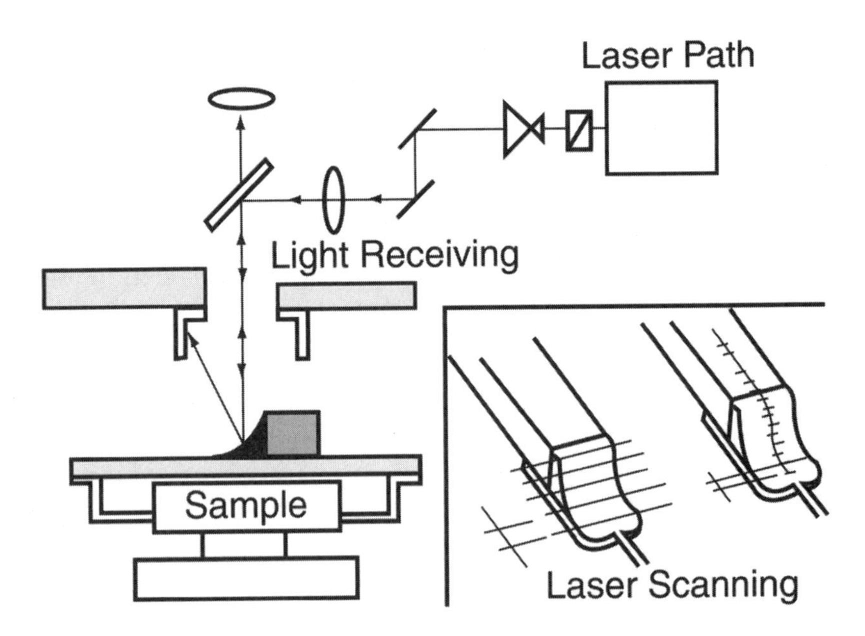

FIGURE 9-1 Principle of laser-type appearance inspection

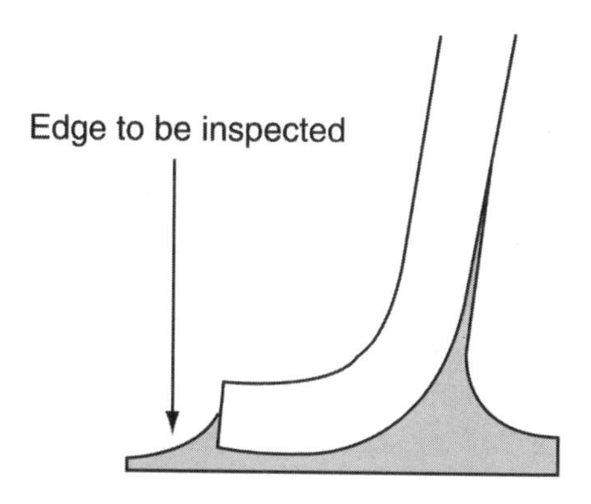

FIGURE 9-2 Cross-section of a soldered joint

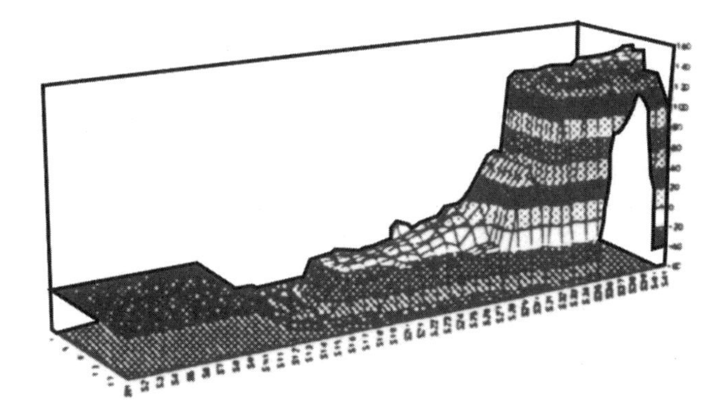

FIGURE 9-3 Image of a good solder joint

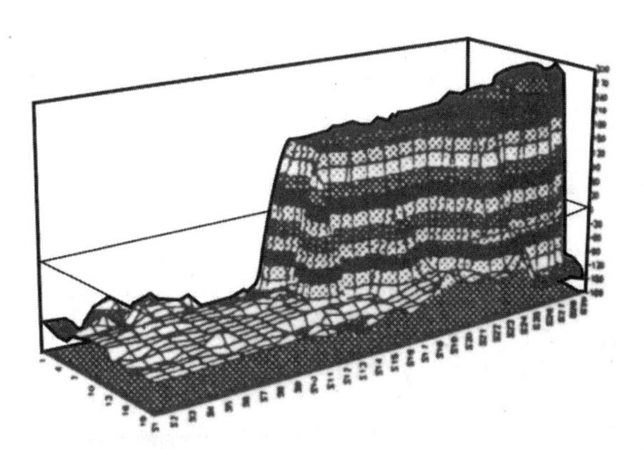

FIGURE 9-4 Image of a bad solder joint

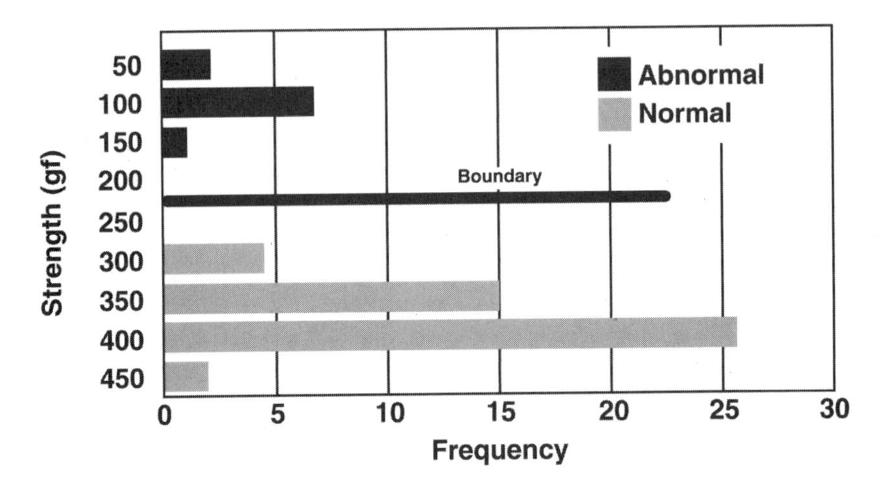

FIGURE 9-5 Soldering strength of good and bad units

make sure that there were no inspection errors, tensile strength tests were conducted in order to compare the two groups. Figure 9-5 shows that the two groups are clearly discriminate.

Two types of results exist:

a) Reflection characteristics: Results showing laser reflection
b) Inspection logic characteristics: Results calculated from laser reflection following the inspection logic

From each type of characteristic, Mahalanobis Distances were calculated, and their discriminating capabilities were compared. The equations of calculation are the same as the equations that we previously described.

9.3 MAHALANOBIS DISTANCE USING INSPECTION-LOGIC CHARACTERISTICS

Among the 130 characteristics from inspection logic, 77 characteristics were used to construct a Mahalanobis Space. Table

9-1 shows the Mahalanobis Distances that were calculated from both the normal and abnormal units. We confirmed from the distribution of Mahalanobis Distances that normal (good) and abnormal (bad) units can be well separated.

Next, we must select the inspection-logic characteristics that are contributing to our study. The 77 characteristics were assigned to an orthogonal array L_{128}. Levels 1 and 2 of each column were assigned as "use" and "not use," respectively. Based on such assignments, we constructed the Mahalanobis Space of

TABLE 9-1 Mahalanobis Distances of Normal and Abnormal Units

Normal				Abnormal
Interval	Frequency	Interval	Frequency	Distance
0	0	5.5	4	37.09
.05	267	6	1	28.24
1	1626	6.5	2	100.39
1.5	423	7	0	37.91
2	175	7.5	2	31.96
2.5	79	8	1	89.38
3	36	8.5	2	89.75
3.5	21	9	2	29.49
4	11	9.5	0	36.64
4.5	8	10	0	49.18
5	5	Next Class	5	25.17
Distance of Next Class			29.20	
			10.12	
			11.92	
			12.69	
			10.39	

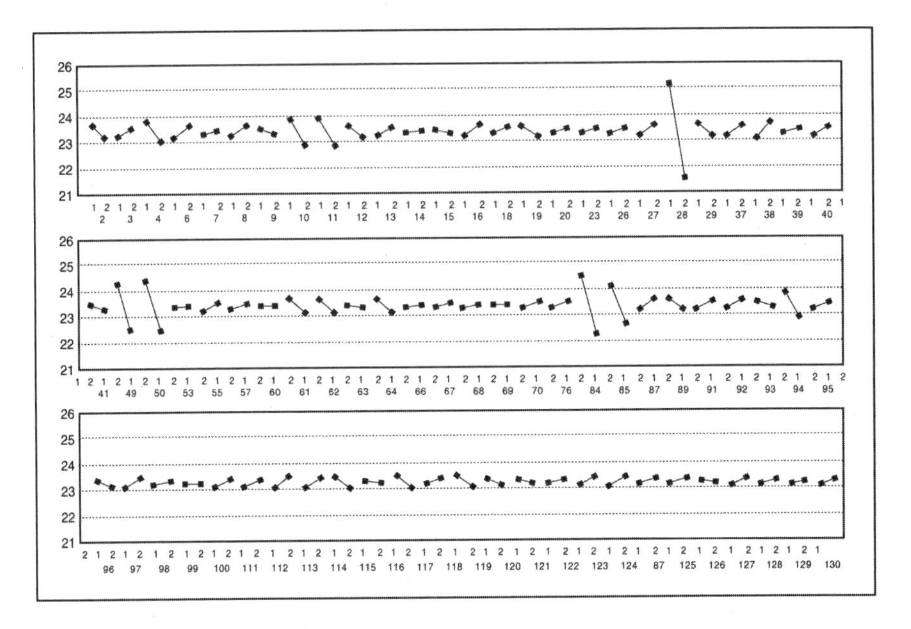

FIGURE 9-6 Factorial effects of inspection-logic characteristics

each row. The abnormal units were assigned outside the orthogonal array. Using the Mahalanobis Space of each row, we calculated the Mahalanobis Distances of abnormal units as D_1^2, D_2^2, . . . , and Dn^2. Then, we calculated the SN ratio (the larger-the-better) of each row. Figure 9-6 shows the effects of inspection-logic characteristics.

Because Level 1 is "used" and Level 2 is "not used," the factorial effect with the left shoulder up shows contributing data, and the right shoulder up shows the data that worsens the precision of inspection. The characteristics that had a 1 db or more difference between the two levels are considered useful and are shown in Table 9-2.

9.4 MAHALANOBIS DISTANCE USING REFLECTION CHARACTERISTICS

Next, we studied the effectiveness of the case that used the reflection data. Similar to the case of inspection-logic charac-

TABLE 9-2 Selected Inspection Using Logic Characteristics

10	convex shape of pad edge
11	average height of solder
28	diffused reflection of rear part after laser scanning
49	flatness of pattern (1)
50	flatness of pattern (2)
84	roundness of solder shape (1)
85	roundness of solder shape (2)
91	convex shape of solder

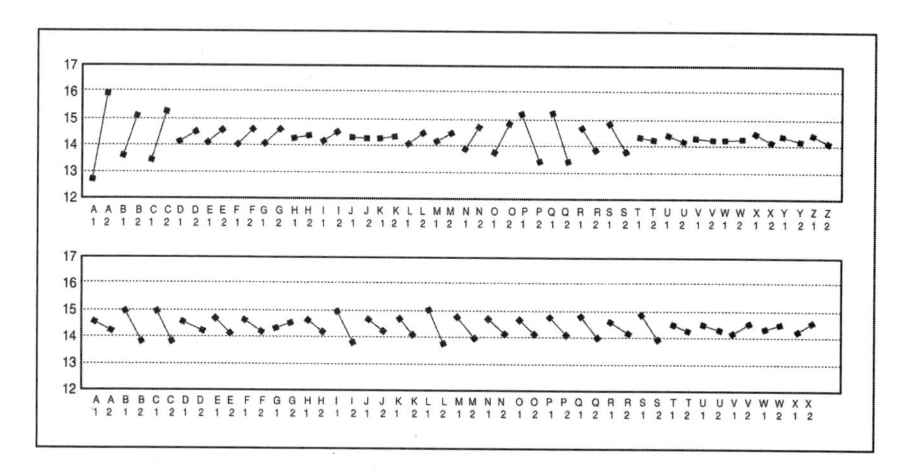

FIGURE 9-7 Factorial effects of reflection characteristics

teristics, 350 normal units and nine abnormal units were taken from the manufacturing process. Fifty characteristics were assigned to the orthogonal array L_{64}. The factorial effects are shown in Figure 9-7.

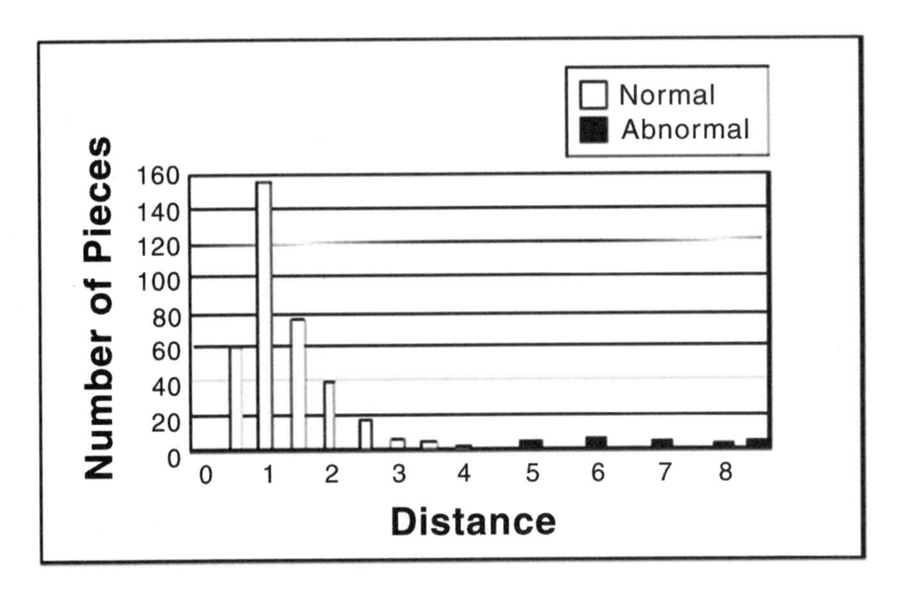

FIGURE 9-8 Mahalanobis Distance using reflection characteristics

From the analysis, 31 characteristics were selected to construct a Mahalanobis Space. Based on this space, the Mahalanobis Distances of normal and abnormal units were calculated as shown in Figure 9-8.

From the figure, we concluded that a good discriminating result can be obtained by using only reflection data.

CHAPTER TEN

FIRE ALARM SYSTEM OPTIMIZATION

10.1 INTRODUCTION

Most of us have experienced false fire alarms in our daily lives. After experiencing a few of these false alarms, we tend to ignore the alarm. Many public facilities, such as hotels, are mandated to install a fire alarm system. If the system issues frequent false alarms, however, the hotel manager tends to cut off the system in another way to avoid inconvenience to customers. These two types of mistakes might one day create catastrophic accidents.[1]

A great deal of variability exists in fire accidents, ranging from small to large. For such abnormal situations, trying to study their distribution is useless. Statistics are calculated based on raw data, where the distribution is practically unknown. The reason why current fire-alarm systems are not working well is because studies have been conducted on fire-detecting capabilities, instead of studying the evaluating method for the group of data collected from the situation of "no fire." Most importantly, we need to collect enough data from "no fire" situations at different times and at various locations. Humans can easily judge whether a fire will erupt by summarizing the conditions of flame and smoke, but this task is not easy for an alarm system.

In this chapter, a fire-detecting system is designed by calculating the Mahalanobis Distance of various situations by using

[1]Kamoshita, Takashi, Kazuta Tabata, Harutoshi Okano, Kazuhito Takahashi, Hiroshi Yano. "Optimization of a Multi-Dimensional Information System Using Mahalanobis Distance," ASI Total Product Development Symposium, 1997.

the Mahalanobis Space, which is constructed from a number of sets of data under "no fire" conditions.

10.2 DATA COLLECTION

In this study, five smoke sensors and five temperature sensors are installed in each of three rooms. Data were collected at different times, as shown in Table 10-1.

Data in this format were collected under normal conditions (i.e., no fire) in three different rooms: A, B, and C. Room A had 40.0 square meters; Room B had 20.5 square meters; and Room C had 19.5 square meters. In each room, there were 10 sensors (i.e., five fire sensors and five smoke sensors). The positioning of sensors in Room A is illustrated in Figure 10-1.

In Figure 10-1, the numbers represent the positioning of five sensor sets. Each set has a smoke sensor (S) and a temperature sensor (T).

TABLE 10-1 Time Intervals for Data Collection

t1	t2	t3	t4	t5
Current	30 sec. Before	60 sec. Before	90 sec. Before	120 sec. Before

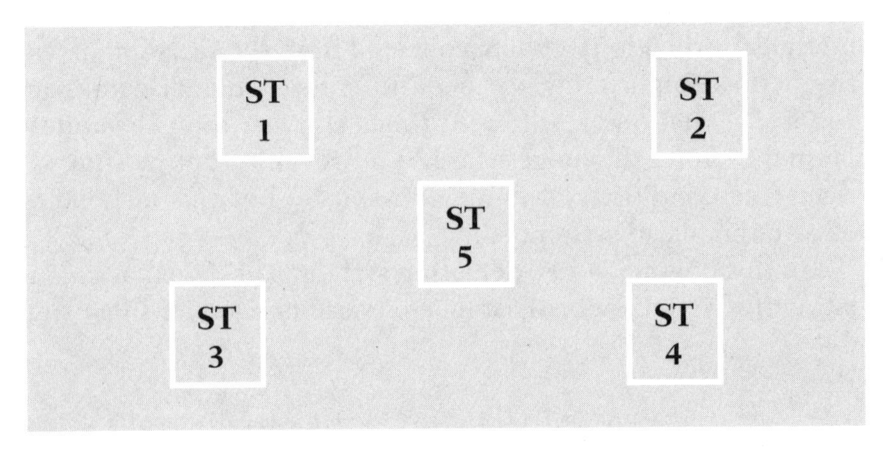

FIGURE 10-1 Positioning of sensors in Room A

The data sets were collected in Room A under normal conditions, with the factors and levels as shown in Table 10-2.

An L_{12} orthogonal array was used to accommodate the five factors. The physical layout is shown in Table 10-3.

Data were collected with windows either open or closed from zero hours to 18 hours at the intervals of six hours for 10 minutes. Therefore, the number of data sets for Room A equals $12 \times 2 \times 4 = 96$. In this equation, 12 corresponds to L_{12}; 2 corresponds to window positions; and 4 corresponds to the data-collection interval (i.e., from zero hours to 18 hours). In addition, 24 sets of data were collected. Hence, the total number of data sets in Room A equaled 120.

Similarly, data sets were obtained in Rooms B and C with the factors and levels shown in Table 10-4.

TABLE 10-2 Factors and Levels for Room A

Factor	Level 1
Temperature (A)	LOW
Mosquito Incense (B)	NO
Cigarette Smoke (C)	NO
Oil Burn (D)	NO
Cooking (E)	NO

TABLE 10-3 L_{12} Orthogonal Array

Test	A	B	C	D	E
1	low	no	no	no	no
2	low	no	no	no	no
3	low	no	yes	yes	yes
4	low	yes	no	yes	yes
5	low	yes	yes	no	yes
6	low	yes	yes	yes	no
7	high	no	yes	yes	no
8	high	no	yes	no	yes
9	high	no	no	yes	yes
10	high	yes	yes	no	no
11	high	yes	no	yes	no
12	high	yes	no	no	yes

TABLE 10-4 Factors and Levels for Rooms B and C

Factor	Level 1	Level 2
Temperature (A)	LOW	HIGH
Mosquito Incense (B)	NO	YES
Cigarette Smoke (C)	NO	YES
Oil Burn (D)	NO	YES

TABLE 10-5 Data Sets

Data Set	Smoke Sensors			Temperature Sensors		
	S_1	....	S_5	T_1	....	T_5
	$t_1 \ldots t_5$		$t_{21} \ldots t_{25}$	$t_{26} \ldots t_{30}$		$t_{46} \ldots t_{50}$
1	Y_{11}	..		..		$Y_{1,50}$
⋮						
322	$Y_{322,1}$	..		..		$Y_{322,50}$

For Rooms B and C, only four columns of L_{12} layout were utilized. Data were collected from zero hours to 18 hours at an interval of six hours for 10 minutes. Therefore, there are $12 \times 4 = 48$ data sets for Rooms B and C. Additionally, 106 data sets were obtained in Room C with ON and OFF positions of injection molding machines situated in Rooms B and C. Therefore, the total number of data sets available equaled $120 + 48 + 48 + 106 = 322$. These data sets were arranged as shown in Table 10-5.

In Table 10-5, $S_1 \ldots S_5$ represents five smoke sensors; $T_1 \ldots T_5$ represents five temperature sensors; and $t_1 \ldots t_{50}$ represents data-collection times, as shown in Table 10-1. You should note that $t_1 \ldots t_{25}$ corresponds to the temperature sensors, and $t_{26} \ldots t_{50}$ corresponds to the smoke sensors.

10.3 CALCULATION OF MAHALANOBIS SPACE

Table 10-6 shows part of the data that were collected from smoke and temperature sensors.

TABLE 10-6 Data Under Normal Conditions (Part)

Temperature Sensor No. 1					
No.	t_1	t_2	t_3	t_4	t_5
	Current	30 sec. before	60 sec. before	90 sec. before	120 sec. before
1	296	296	296	296	296
2	296	296	293	293	293
3	0	0	0	0	0
4	0	0	0	0	0
5	0	0	0	0	0
...	...	...	...	...	...

The mean, denoted by *m*, and the standard deviation, denoted by σ, are calculated as shown in Table 10-7. For example,

$$y_{11} = \frac{Y_{11} - m_1}{\sigma_1} = \frac{0 - 0.27}{0.88} = -0.311$$

The results of normalized data are shown in Table 10-8. Next, we calculate the correlation matrix, denoted by R. The correlation coefficient, r_{11}, is calculated as follows:

TABLE 10-7 Mean and Standard Deviation

Item	1	2	3	4	....	50
Mean	0.27	0.28	0.29	0.30	....	301.31
Standard Deviation	0.88	0.91	1.08	1.24	....	3.48

unit: smoke sensor: Concentration (%)
temperature sensor: °k (Kelvin)

TABLE 10-8 Normalized Data

No.	Smoke Sensor No. 1					...	Temperature Sensor No. 5	
	t_1	t_2	t_3	t_4	t_5	...	t_{49}	t_{50}
1	-0.311	-0.303	-0.261	-0.242	-0.201	...	-1.443	-1.527
2	-0.311	-0.303	-0.261	-0.242	-0.201	...	-2.265	-2.300
3	-0.311	-0.303	-0.261	-0.242	-0.201	...	-1.442	-1.527
...	...	...	...	...	...	... ...		

$$r_{11} = \frac{1}{322}[(- 0.311) \times (- 0.311) + (- 0.311)$$

$$\times (- 0.311) + ... + (- 0.311) \times (- 0.311)]$$

$$= 1.00 \tag{10.1}$$

The correlation matrix will have the following structure:

$$R = \begin{bmatrix} 1 & Y_{12} & \cdots & Y_{1.50} \\ Y_{21} & 1 & \cdots & Y_{2.50} \\ \cdots & & \cdot\cdot & \cdot\cdot \\ Y_{50.1} & \cdot\cdot & \cdots & 1 \end{bmatrix} = \begin{bmatrix} 1.00 & 0.95 & \cdots & 0.10 \\ 0.95 & 1.00 & \cdots & 0.08 \\ \cdot\cdot & \cdots & & \\ 0.10 & 0.08 & \cdots & 1.00 \end{bmatrix} \tag{10.2}$$

The inverse of R, denoted by $A = R^{-1}$, is shown as follows:

$$A = R^{-1} = \begin{bmatrix} a_{11} & a_{12} & \cdots & a_{1.50} \\ a_{21} & a_{22} & \cdots & a_{2.50} \\ \cdots\cdots & & & \\ a_{50.1} & a_{50.2} & \cdots & a_{50.50} \end{bmatrix} = \begin{bmatrix} 26.99 & -21.23 & \cdots & 1.58 \\ -21.23 & 36.46 & \cdots & Y_{2.50} \\ \cdots & & & \\ 1.58 & -2.08 & \cdots & 17.35 \end{bmatrix} \tag{10.3}$$

This inverse matrix is called Mahalanobis Space.

10.4 CALCULATION OF MAHALANOBIS DISTANCE

Using the elements of the inverse matrix, the Mahalanobis Distance of the No. 1 set, denoted by D_1^2, is calculated as follows:

$$D_1^2 = \frac{1}{50}[26.99 \times (-0.311) \times (-0.311) + \ldots$$

$$+ 17.35 \times (-1.527) \times (-1.527)]$$

$$= 1.11 \tag{10.4}$$

TABLE 10-9 Mahalanobis Distance of the Normal Condition

1.11 1.57 0.25 1.39 1.21 1.17 .86 1.36 1.05 1.55 1.08 .49 1.05 1.4 0.99 1.41 0.49
1.32 1.05 1.49 0.36 0.72 0.21 1.00 0.98 0.88 0.38 0.97 1.22 0.84 0.25 1.47 0.72 0.94
0.13 0.74 0.47 0.45 0.18 0.23 0.23 0.52 0.26 0.38 0.34 0.47 0.53 0.7 0.39
0.19 0.42 0.17 0.12 0.12 0.12 0.12 0.12 1.66 1.04 0.75 0.68 1.17 1.32 0.62 0.69 0.75
0.18 0.18 1.32 1.78 1.66 1.09 0.61 0.33 0.34 1.16 1.48 0.55 1.47 1.83 0.53 0.70 0.30
1.05 0.39 1.92 0.33 0.79 0.62 0.65 0.74 2.01 0.42 0.39 0.37 1.38 1.40 0.89 0.29 2.36
1.43 1.68 0.67 0.58 1.14 0.64 0.64 0.33 0.75 1.19 0.20 1.16 0.62 0.95 1.39 0.76 1.27
0.82 0.15 1.40 0.99 0.84 0.55 0.75 1.34 1.00 5.27 0.67 0.72 0.14 0.72 0.08 0.51 0.13
0.21 0.05 0.54 0.05 0.64 0.58 1.28 0.46 0.95 0.44 0.04 0.15 0.18 0.05 0.84 0.08 0.75
0.60 0.86 0.29 1.08 0.15 0.35 0.28 0.62 0.60 0.52 0.28 0.13 1.72 1.01 0.61 0.60 4.37
0.50 0.16 0.14 0.07 0.72 0.43 0.04 0.29 0.21 0.95 2.18 0.56 0.27 0.45 1.65 0.67 2.79
4.34 0.22 0.96 3.48 3.02 3.43 0.45 0.54 0.40 2.21 0.50 1.30 2.90 0.26 0.41 2.29 2.07
3.15 0.54 0.36 1.03 1.53 0.15 3.34
0.69 0.46 0.45 2.91 3.54 2.59 0.20 0.15 0.55 1.63 0.44 1.04 0.87 0.26 0.47 3.45 1.60
3.38 0.34 0.18 0.73 3.17 0.43 2.07 2.65 0.44 1.06 1.94 1.06 3.54 0.27
0.13 0.55 4.67 0.11 5.14 2.58 0.29 0.29 4.36 1.95 3.54 0.90 0.54 0.37 3.69 0.41 3.88
2.21 0.25 0.74 2.29 2.70 2.31 0.44 0.09 0.59 3.75 4.26 3.43 2.39 0.78 0.30 0.50 0.56
1.16 0.88 1.05 0.87 0.78 0.65 0.08 0.60 0.65 0.30 0.19 0.11 0.28 0.11 0.04 0.96 0.57
1.11 0.71 0.56 1.36 0.38 0.41 0.73 0.34 0.11 0.1 0.31 0.13 0.05 0.05 0.05 0.05 0.05
0.59 0.29 0.05 0.05

TABLE 10-10 D^2 Values for Case 1

Condition	D^2
a	95.1
b	9.05
c	166.73

The results of D^2 of the normal condition are shown in Table 10-9.

The average value of D^2 in Table 10-9 is $D^2 = 1.005$.

After constructing the Mahalanobis Space, the room conditions were changed in the following cases with the same experimental setup.

CASE 1: GENERATION OF ARTIFICIAL FIRE CONDITIONS. In Room A, food was cooked and smoke was purposely generated.

Researchers collected three sets of data: a, b, and c. Table 10-10 shows the D^2 of each set of data.

CASE 2: GENERATION OF DATA UNDER NORMAL CONDITIONS. Researchers collected three sets of data—a, b, and c—under normal conditions. These data are not included in the data that generated the Mahalanobis Space. The D^2 values are still within the normal range and are shown in Table 10-11.

CASE 3: GENERATION OF ARTIFICIAL SMOKE BY BARBECUE COOKING. Researchers used a gas range to create a barbecue

TABLE 10-11 D^2 Values for Case 2

Condition	D^2
a	1.30
b	1.76
c	1.87

TABLE 10-12 D^2 Values for Case 3

Condition	D^2
a	4.12
b	4.26
c	3.43

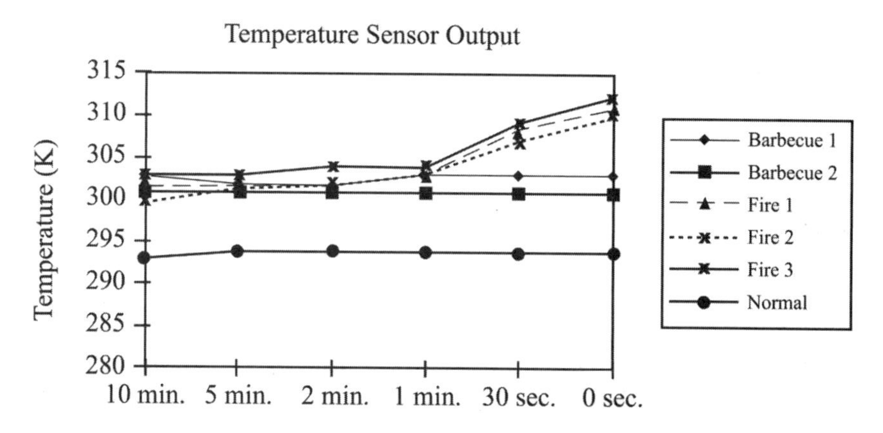

FIGURE 10-2 Temperature change by environmental changes

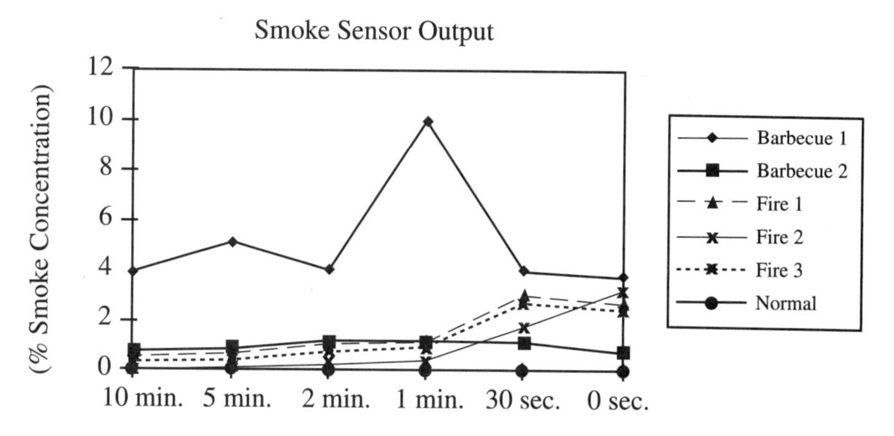

FIGURE 10-3 Smoke concentration by environmental changes

in Room A and collected three sets of data: a, b, and c. The D^2 values shown in Table 10-12 were greater than normal conditions, but they were smaller than the values of a fire situation.

Figures 10-2 and 10-3 show the output response of sensors according to environmental changes. Note that it takes a long time for temperature sensors by themselves to detect a fire, and a smoke sensor by itself cannot separate a fire from a barbecue.

10.5 SELECTION OF SENSORS

As described previously, there are five smoke sensors and fire temperature sensors in the study. In order to reduce the cost of the fire alarm system, researchers tried to reduce the number of sensors without sacrificing their sensing capabilities.

STEP 1

Of the 10 sensors, the smoke sensor and the temperature sensor in the middle of the room were kept and used continually. The remaining eight sensors (represented by letters A through G in Figure 10-4 and Table 10-13) were assigned to the L_{12}

TABLE 10-13 Factors and Levels for the Experiment

Factor	Level 1	Level 2
1	Use A	Don't Use A
2	Use B	Don't Use B
3	Use C	Don't Use C
4	Use D	Don't Use D
5	Use E	Don't Use E
6	Use F	Don't Use F
7	Use G	Don't Use G
8	Use H	Don't Use H

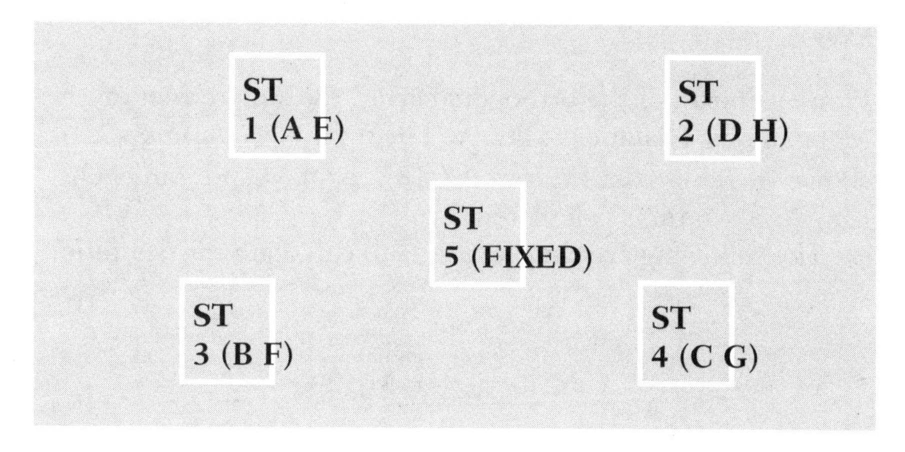

FIGURE 10-4 Sensor arrangement for experiment

orthogonal array. These sensors were considered control factors for the selection of either Level 1 or Level 2.

STEP 2

Researchers collected nine sets of data under normal conditions (no fire) and nine sets of data after a fire by using the Mahalanobis Distance from each set of data as the signal-factor levels. For example,

$$M_1 = \sqrt{D_1^2} = 0.53, M_2 = \sqrt{D_2^2} = 0.69, ..., M_{18} = \sqrt{D_{18}^2} = 79.81$$

STEP 3

From each of the 12 runs, a Mahalanobis Space is constructed from the sensors that are assigned as Level 2 ("use"), together with the two sensors in the middle of the room.

STEP 4

From the 18 sets of data that include nine sets with "no fire" conditions and nine sets with "fire" conditions, 18 Mahalanobis Distances, D^2, are calculated from the Mahalanobis Space of that run. (The run is constructed in Step 3.) In total, there will be 12 × 18 Mahalanobis Distances, or 216.

STEP 5

From each run of the orthogonal array, the square root of Mahalanobis Distance is denoted by u_1, u_2, . . . , and u_{18}, as shown in Table 10-14. From the 18 sets of M and u in each run, an SN ratio is calculated.

The following are the equations to calculate the SN ratio:

$$r = M_1^2 + \dots + M_{18}^2 \tag{10.5}$$

$$\beta = \frac{M_1u_1 + \dots + M_{18}u_{18}}{r} \tag{10.6}$$

$$S_\beta = \frac{(M_1u_1 + \dots + M_{18}u_{18})^2}{r} \tag{10.7}$$

$$S_T = u_1^2 + u_2^2 + \dots + u_{18}^2 \tag{10.8}$$

$$S_e = S_T - S_\beta \tag{10.9}$$

$$V_e = \frac{S_e}{n - 1} \tag{10.10}$$

$$\eta = 10\log\frac{\frac{1}{r}(S_\beta - V_e)}{V_e} \tag{10.11}$$

TABLE 10-14 Results of No. 1

Signal	M_1	M_2	....	M_{18}
	0.53	0.69	....	79.81
u	1.03	1.22	....	17.82

$$S = 10\log\frac{1}{r}(S_\beta - V_e) \tag{10.12}$$

The SN ratio of No. 1 is calculated as follows:

$$r = 0.53^2 + 0.69^2 + \ldots + 79.81^2 = 13626.92 \tag{10.13}$$

$$\beta = \frac{0.53 \times 1.03 + \ldots + 79.81 \times 17.82}{13626.92} = 0.303 \tag{10.14}$$

$$S_\beta = \frac{(0.53 \times 1.03 + \ldots + 79.81 \times 17.82)^2}{13626.92} = 1251.11 \tag{10.15}$$

$$S_T = 1.03^2 + 1.22^2 + \ldots + 17.82^2 = 2172.85 \tag{10.16}$$

$$S_e = 2172.85 - 1251.11 = 921.74 \tag{10.17}$$

$$V_e = \frac{921.74}{18 - 1} = 54.22 \tag{10.18}$$

$$\eta = 10\log\frac{\frac{1}{13626.92}(1251.11 - 54.22)}{54.22} = -27.91(\text{db}) \tag{10.19}$$

$$S = 10\log\frac{1}{13626.92}(1251.11 - 54.22) = -0.56(\text{db}) \tag{10.20}$$

Figure 10-5 shows the response graph of eight sensors.

From Figure 10-3, we see that smoke sensors No. 1 and No. 4—and also temperature sensors No. 1, 2, and 4—are effective. To reduce cost, researchers decided to use smoke sensors No. 1 and No. 4 together with the basic two sensors (one smoke sensor and one temperature sensor in the middle of the room), as shown in Figure 10-6. Figure 10-7 shows the differences before and after optimization.

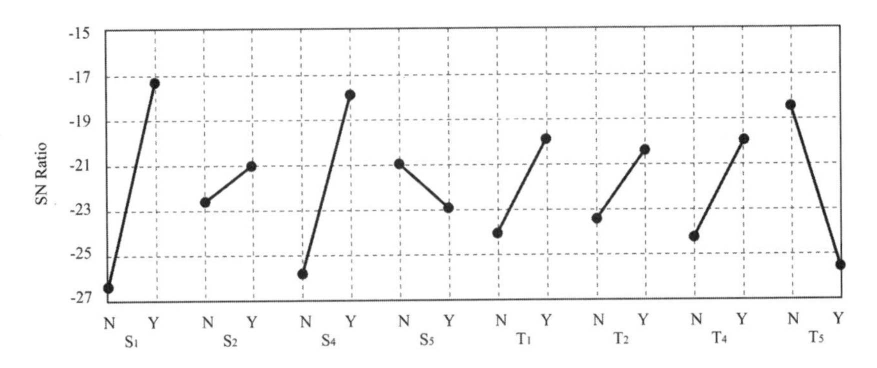

FIGURE 10-5 Response graph

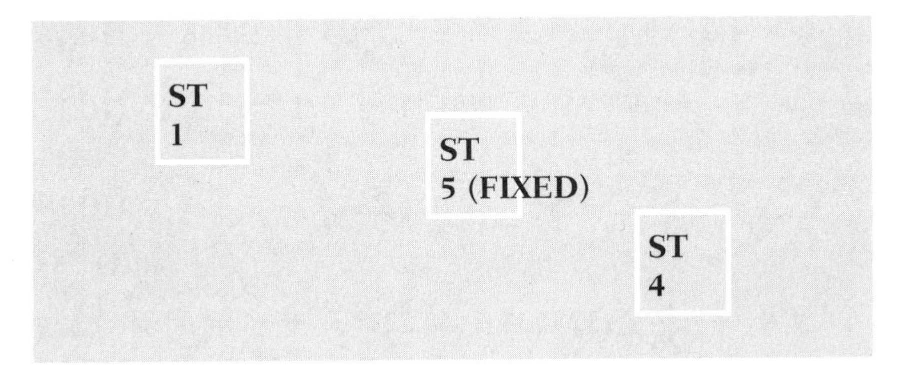

FIGURE 10-6 Sensor arrangement for effective functioning of a fire-alarm system

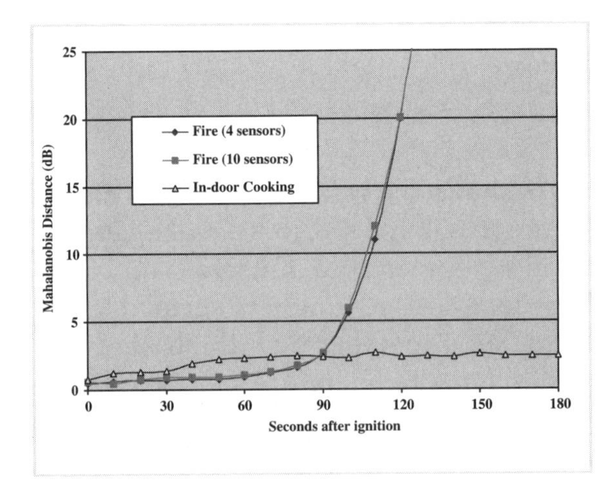

FIGURE 10-7 Result from the optimum diagnosis system using Mahalanobis Distance

SECTION FOUR

THE CHEMICAL INDUSTRY

CHAPTER ELEVEN

DIAGNOSIS OF PHOTOGRAPHIC-PROCESSING SOLUTION

11.1 INTRODUCTION

In photography, photographers and technicians use many analysis items to assess the condition of processing solutions. In order to know the quality of processing solutions on the market, these photographers have traditionally assessed products based on experience by observing the differences between the constituents of the solution to be assessed and the constituents of the new solution. They should also make assessments based on the type of processing equipment used in the stores, however. Also, the correlations of this equipment must be considered. We applied the MTS method in the following study in order to quantify such assessments.[1]

11.2 PROCESSING OF PHOTO-SENSITIVE MATERIALS

Photographic film contains silver halide photo-sensitive materials. After light exposure through cameras, the film is processed

[1]Kanazana, Yukihiko, Shinzou Kishimoto, Jun Okamoto (Fuji Photo Film Co). "An Application of MTS for Qualitative Diagnosis of Photographic Processing Solution," Quality Engineering Forum Symposium, 1998.

in four steps: developing, fixing, washing, and drying. In the development step, the photo-sensitive particles that formed the latent image after exposure are selectively reduced (developed) by using the developing solution. In the fixing step, the fixer solution dissolves and removes unreacted particles.

The developing solution is composed of a primary developing agent, preservative, and development inhibitor. These ingredients vary in a complex manner and mutually interact, depending on the processing conditions. In this study, we used MTS to diagnose silver halide photo-sensitive materials in the processing stage, including the developer and fixer solutions.

The objectives of MTS application include the following:

1) To quantitatively diagnose solutions by using Mahalanobis Space as a base to substitute the method of qualitative evaluation, based on the comparison with a new solution
(2) To diagnose solutions by using one scale and considering the correlations between multi-dimensional information
(3) To select the items that contribute to Mahalanobis Distance

11.3 SELECTION OF THE BASE SPACE

In a previous study, we constructed the base space with both developer and fixer solutions. But diagnoses were made occasionally for either developer solutions or fixer solutions only. Therefore, we constructed the space from each solution separately in this study.

For the construction of the base space, we selected the developer solutions from stores in the market that had no problems with photographic quality or other aspects. We collected one hundred sets of data from those stores. The characteristics (items) included the composition and physical properties of the solutions as well as the treatment conditions, as shown in Table 11-1.

TABLE 11-1 Mahalanobis Space Setting

Item / Store	Condition				Solution	
	Machine	Quantity	Treatment Condition 1	Treatment Condition 2	Property	Constituent
	A, B, C ...	A, B, C ...	A, B, C ...	A, B, C ...	A, B, C ...	A, B, C ...
1 2 . . . 100						

TABLE 11-2 Mahalanobis Distance of Solutions that Received Complaints

Store	Mahalanobis Distance	Traditional Assessment
A	1.1	Absolutely normal solution. Possibly due to other causes.
B	2.2	Normal solution.
C	2.8	Slightly abnormal solution.
D	3.5	Slightly abnormal solution (slightly oxidized by air).
E	4.5	Abnormal solution (concentrated).
F	7.0	Abnormal solution (concentrated).
G	31.6	Abnormal solution (highly concentrated).

11.4 MAHALANOBIS DISTANCE OF REJECTED SOLUTIONS AND THEIR PHOTOGRAPHIC QUALITY

Our past records show that insufficient maximum density was the most common cause of photographic complaints. From the data that we collected from the stores that had the most complaints, we calculated Mahalanobis Distances by using the base space mentioned previously, as shown in Table 11-2. In the table, we made a comparison between Mahalanobis Distance and traditional assessments.

We see from the table that among these solutions, the ones that were judged as normal by traditional assessments had small Mahalanobis Distances. The solutions that were judged as defective by traditional assessment had large Mahalanobis Distances. These facts indicate a good correlation between traditional assessments and the Mahalanobis Distance approach.

To study the correlation between Mahalanobis Distance and photographic properties (density), we performed running tests in order to cause the deterioration of processing solutions by using commercial methods. Table 11-3 and Figure 11-1 show the results.

We see from the table and the figure that there is an approximate correlation between Mahalanobis Distance and photographic quality. From the correlation, we can roughly determine the point at which problems occur. A distance of 2.5 can be used as a threshold.

TABLE 11-3 Correspondence between Mahalanobis Distance and Photographic Quality

Solution	Mahalanobis Distance	Photographic Quality (max. concentration)
New	1.0	2.25
Intermittent Sampling 1	1.8	5.3
Intermittent Sampling 2	2.2	4.8
Intermittent Sampling 3	2.2	4.7
Intermittent Sampling 4	3.3	4.6
Final Solution (equilibrium state)	4.6	4.35

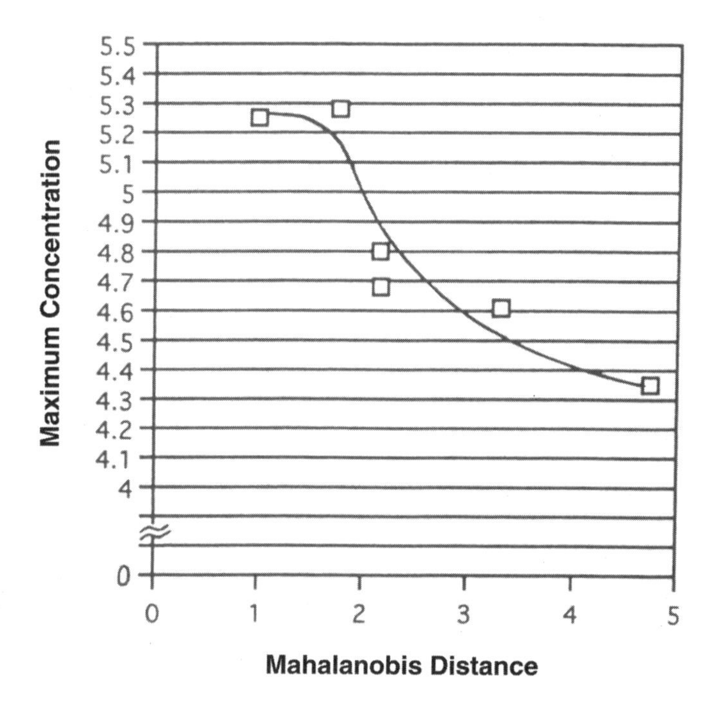

FIGURE 11-1 Correspondence between Mahalanobis Distance and photographic quality (concentration)

11.5 FACTORIAL EFFECTS

In order to find the causes of these users' complaints, we can observe the factorial effects on Mahalanobis Distance. Figure 11-2 shows the factorial effects of a user who complained (user G) by using Mahalanobis Distance for analysis. Figure 11-3 shows the results from all users who complained by analyzing the larger-the-better SN ratios.

11.6 DISCUSSIONS

All new solutions have the same constituents. The solutions that we took from users for the construction of the base space (100 sets of data) deteriorated to a certain extent. We collected the data in this manner because there are few cases of totally new solutions on the market.

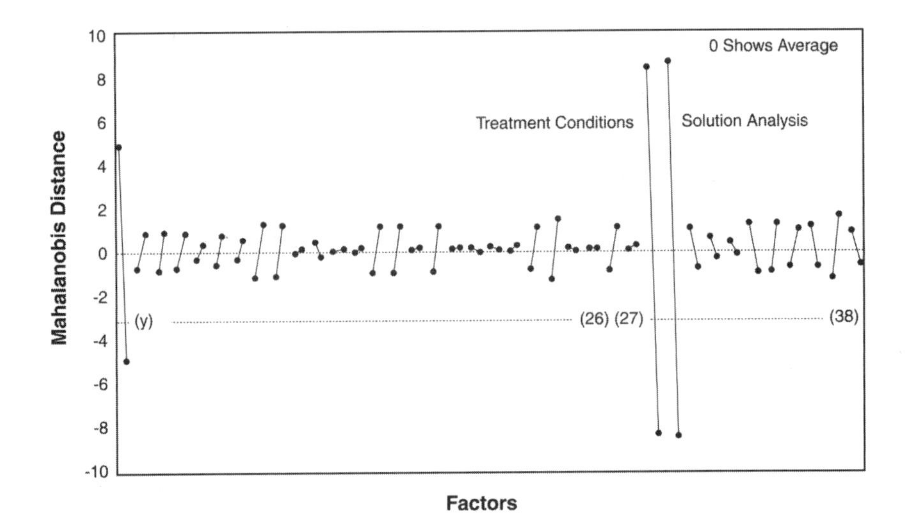

FIGURE 11-2 Factorial effects of a store that received complaints (Store G)

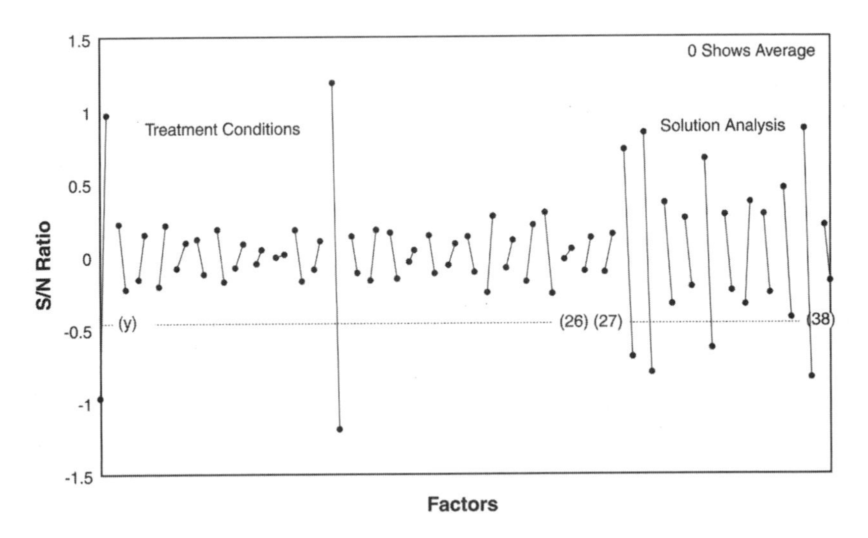

FIGURE 11-3 Factorial effects of all stores that received complaints

Traditionally, assessments have been made based on new solutions, because new solutions never create problems. The Mahalanobis Distance of a new solution, however—calculated by using the base space (constructed from the solutions on the market, i.e., partially deteriorated)—became a large value.

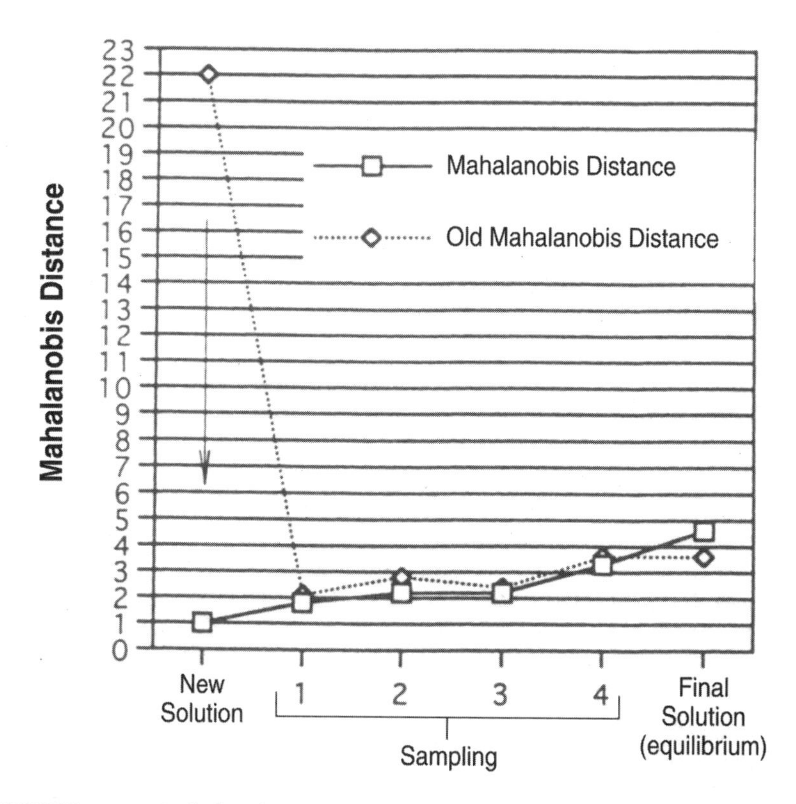

FIGURE 11-4 Mahalanobis Distance changes of old and new solutions during running tests

After discussion, we decided to construct a new base space by putting the fresh solutions together with the solutions that were collected from 100 users, as shown in Table 3-1. Figure 11-4 compares the Mahalanobis Distance changes of old and new spaces during the running tests.

We can see from Figure 11-4 that the Mahalanobis Distance of a new solution is about 1, and the distance smoothly increases when the solution deteriorates.

CHAPTER TWELVE

PATTERN RECOGNITION FOR INFRARED ABSORPTION SPECTRUM ANALYSIS

12.1 INTRODUCTION

Infrared absorption spectrum analysis is conducted in order to measure the absorption of light at different frequencies when infrared rays pass through a sample. This technique is used to identify chemicals and is included in the *Japan Pharmacopoeia*. Because of its simple measuring procedures, low measurement cost, and abundant amount of information, this method has been highly recommended and is considered important. The results are normally shown in graph format, with the wavelength as the abscissa and the transmittance as the ordinate.

According to the procedures given by the *Japan Pharmacopoeia*, the spectrum of a sample is compared with the spectrum of the substance to be identified. The chemical's identity is confirmed when both spectrums absorb the same intensity at the same frequency. Because of measurement errors, the spectrum from the same sample varies more or less every time. For this problem, the *Japan Pharmacopoeia* assumes no errors when measured by an analyst who has reached a certain skill level. In this study, Mahalanobis Distance was used to quantify the degree of agreement.[1]

[1]Asahara, Hatsuki, Yuichiro Ota, Shuichi Takeda, Masahiko Marumoto, Toshitsugu Sato (Tsumura & Co.). "Pattern Recognition for Infrared Absorption of Spectrum," Quality Engineering Forum Symposium, 1998.

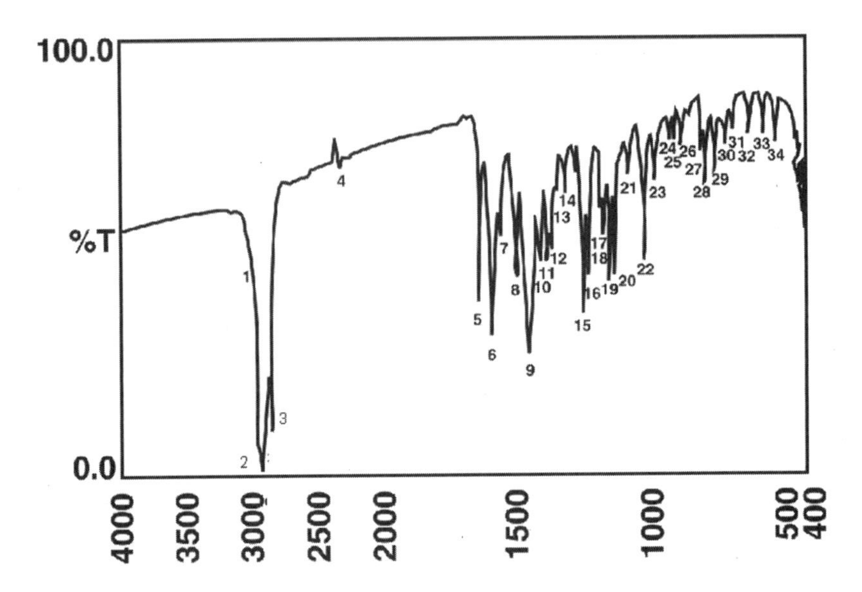

FIGURE 12-1 Infrared spectrum of a normal sample

12.2 EXPERIMENT

For the experiment, we used a chemical that was under development in our company. This material was of a crystal polymorphism type—the type with the same chemical constitution but different in physical or chemical properties.

Figure 12-1 shows the infrared spectrum of a chemical that has the crystal form that we are looking for, and Figure 12-2 shows the spectrum whose crystal form is out of specification. These two spectra look alike and are easy to misidentify, even by an experienced analyst.

Figure 12-3 shows the principle of infrared spectrum measuring equipment made by the Nippon Bunko Company (Model FT/IR-430).

From each normal sample, we took one to three measurements and collected a total of 202 sets of data. From the abnormal samples, we took 15 measurements.

12.3 SELECTION OF CHARACTERISTICS

In pattern recognition, we must consider correlation and variability between peaks. Therefore, wave numbers were consid-

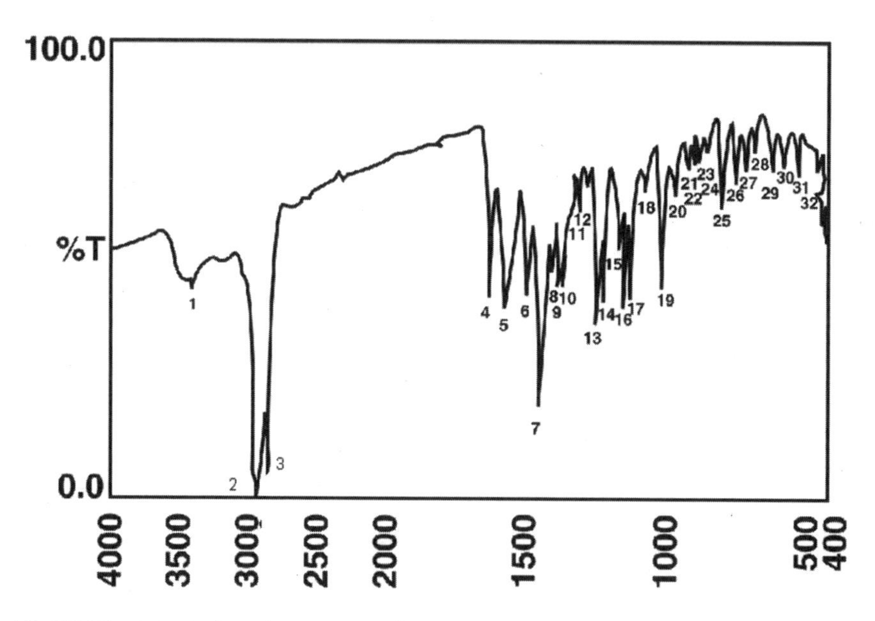

FIGURE 12-2 Infrared spectrum of an abnormal sample

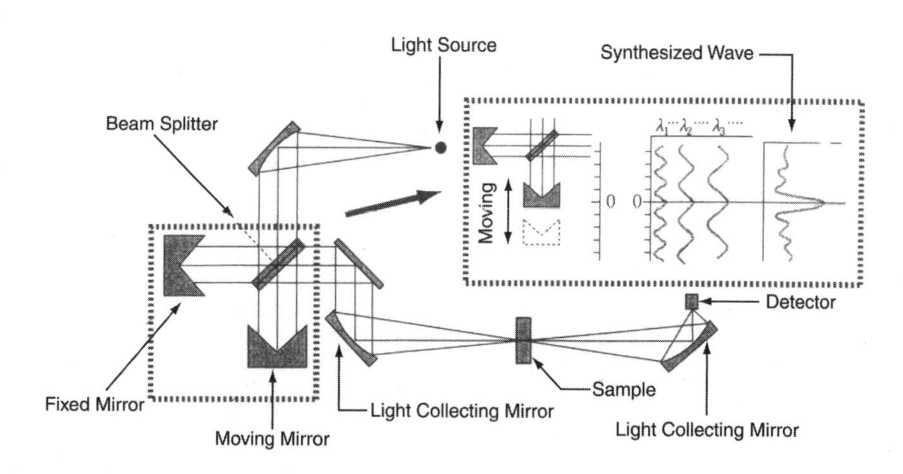

FIGURE 12-3 Principle of infrared absorption spectrum measurement

ered as characteristics, and the absorption rate was used as the output, as shown in Figure 12-4 and Table 12-1. From 202 sets of data, 187 sets were used to construct a base space after randomly discarding 15 sets. Fifteen abnormal samples were also measured in the same way to collect 15 sets of data.

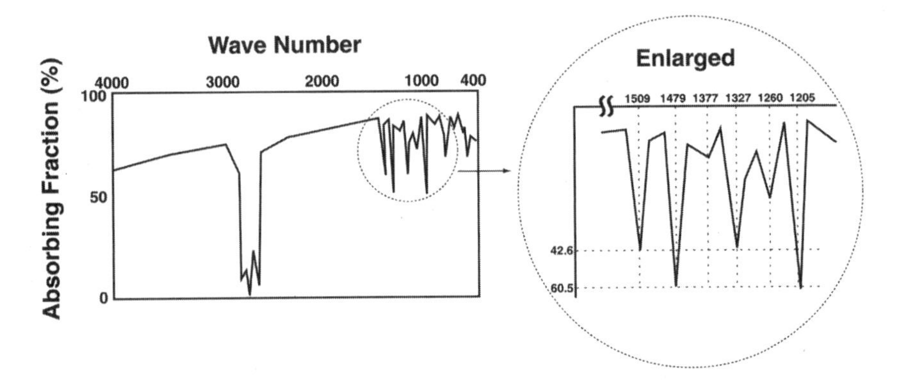

FIGURE 12-4 Characteristics and outputs

TABLE 12-1 Absorbing Fraction and Wave Number

	Item (Wave Number)									
	3253	...	1509	1479	1377	1327	1260	1205	...	718
1	55.7	...	42.6	60.5	10.2	42.3	25.5	56.9	...	55.4
2	64.7	...	45.5	59.0	11.1	40.5	30.0	65.7	...	62.3
.	.	.	.	.	.	.	.	.	.	.
.	.	.	.	.	.	.	.	.	.	.
.	.	.	.	.	.	.	.	.	.	.
202	42.9	...	43.0	53.2	13.3	53.4	28.0	42.8	...	44.9

12.4 RESULTS

During the process of calculation, many pairs were found to have a high correlation coefficient between peak combinations. The variation of infrared absorption spectrum is caused by the amount of the sample or the layer length, which affects the intensity of peaks. The variation of the ratio between peaks is rather small, however. That is, once the absorption rate of the first peak is determined, the magnitudes of the second or third peaks are almost determined. That was probably the reason for having high correlation coefficients. Therefore, those peak combinations that gave the correlation coefficients a value that was higher than 0.99 were removed from the characteristics. This issue will be discussed later.

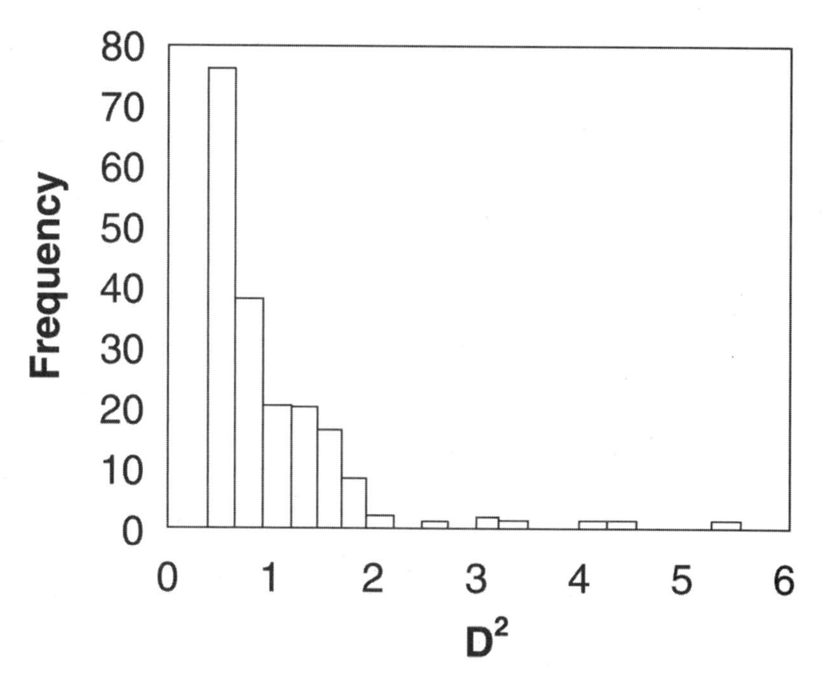

FIGURE 12-5 Histogram of Mahalanobis Distance from the data that were used as the base space

Figure 12-5 shows the distribution of Mahalanobis Distance of 187 normal data sets. From the figure, we see that the Mahalanobis Distances from the samples that have the normal crystal form are within a range of 0.4 to 5.5.

From the 15 sets of data, which were arbitrarily separated from the normal samples, their Mahalanobis Distances were calculated by using the base space that was constructed from the 187 sets of normal samples. Also, the distances were calculated from the 15 sets of abnormal samples. Table 12-2 shows the results. The distances of the two groups are quite different, indicating a good separation.

In this study, we noticed strong correlation coefficients. For this problem, we removed the highly correlated items (characteristics). As a result, the amount of information was reduced. In this experiment, only one piece of measuring equipment was

TABLE 12-2 Mahalanobis Distance

Normal	Abnormal
0.40	570.6
0.50	579.1
0.60	575.9
0.58	570.4
1.07	571.3
1.30	572.8
1.36	525.1
1.70	529.1
0.59	525.9
0.46	494.1
2.83	493.0
0.83	493.0
0.48	359.9
1.81	361.0
1.64	359.7

used in the entire experiment. Also, the same samples were repeatedly measured, which might have resulted in a smaller base-space estimation. In future studies, we should use several pieces of measuring equipment.

SECTION FIVE

THE SPACE INDUSTRY

CHAPTER THIRTEEN

FAULT ANALYSIS

13.1 INTRODUCTION

Autonomous guidance and control systems in current and future space vehicles must be provided with fault-tolerance technology. The starting point of the technology is fault diagnosis; i.e., fault detection and identification.

The *Fault Diagnosis Program* (FDP), a case study for a future Japanese space vehicle, determines the condition of the vehicle and decides whether the on-board sensors work well or not. Also, FDP indicates which sensor has a fault, based on the information from the state estimation errors calculated by observers. In this chapter, we examine the applicability of MTS to FDP of the simplified sensor model.[1]

First, we created a standard space based on residuals, which are calculated from state estimation errors that observers make. We tried the fault detection and identification under that standard space, but the results could not identify some fault cases well.

We re-examined the characteristics of the standard space by changing them from "residuals" into "the estimation errors," and then we examined the possibility of fault detection and identification again. Furthermore, we investigated the methods to reduce the influence of high correlative items in the correlation matrix.

[1]Matsuda, Rika, Yoshiki Ikeda (Mitsubishi Space Software Co.) Takashi Kamoshita (National Research Laboratory of Metrology), Kazuyuki Touhara (National Space Development Agency of Japan). "Application of Mahalanobis Taguchi System to the Fault Diagnosis Program," ASI Symposium, 1998.

13.2 OUTLINE OF THE FAULT DIAGNOSIS PROGRAM (FDP)

The outline and the block diagram of our FDP are shown in Figures 13-1 and 13-2, respectively. Figure 13-2 shows the functions of the dynamics model, the sensor fault simulator, the controller model, and the actuator model as the simulation environment of the FDP. To simplify the study, we assume that the dynamics model outputs can be directly observed as sensor outputs.

Objective System :	HOPE lateral dynamics in the reentry phase
Observer :	It estimates the states from sensor outputs and actuator commands, and generates a "Residual" by using the estimation errors (i.e. difference between sensor outputs and the estimated states).
Fault Decision :	Deciding faults based on the detection rules.

HOPE :
H-II Orbiting Plane, a Japanese unmanned reentry vehicle.

FIGURE 13-1 Outline of the program

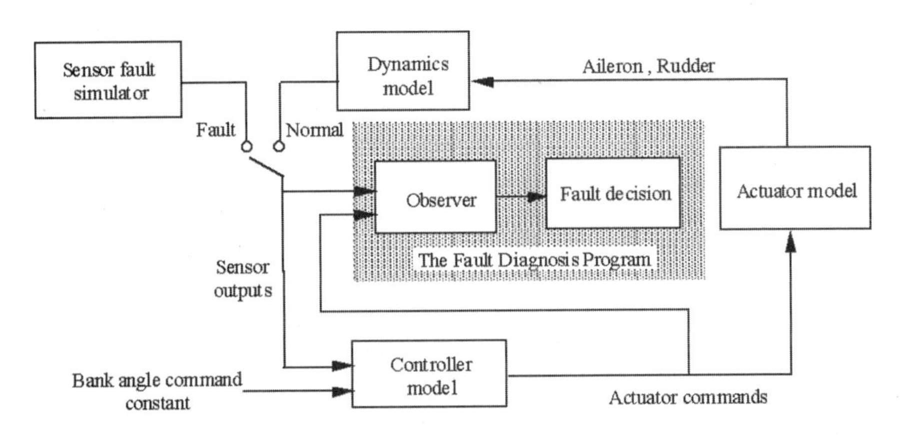

FIGURE 13-2 The block diagram of FDP

13.3 OBSERVER AND RESIDUAL

If the information about the states of the objective system cannot be directly obtained—but if the linearized mathematical model of the system is known—then you can construct the software system that is used to estimate the states by using detectable inputs and outputs. This estimation system is called "observer," and the observer formulation is shown in Figure 13-3.

In this case study, the inputs to the observers are sensor outputs and actuator commands, then the observers estimate the states of *H-II Orbiting Plane* (HOPE, a Japanese unmanned reentry vehicle) lateral dynamics. The sensor model is simplified so that the dynamics model outputs—the bank angle (σ), the side-slip angle (β), the roll rate (p), and the yaw rate (r)—can be directly observed as sensor outputs.

The observer scheme in this case study is shown in Figure 13-4. The scheme consists of four observers, and each observer receives three sensor outputs as inputs. When one sensor fails,

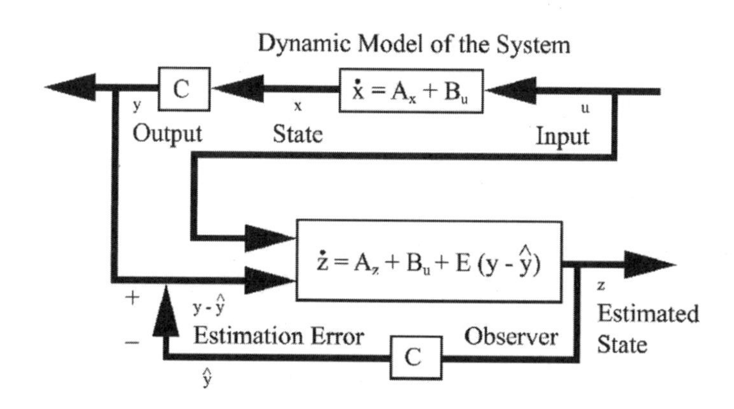

$$\dot{z} = (A - EC)z + B_u + EC_x$$

$$\dot{z} - \dot{x} = (A - EC)(z - x)$$

$$\frac{d}{dt}(z - x) = (A - EC)(z - x)$$

$$z = x + e^{(A-EC)t}(z_0 - x_0)$$

If **A - EC** is stable, **x** can be estimated by **z**.

FIGURE 13-3 The observer formation

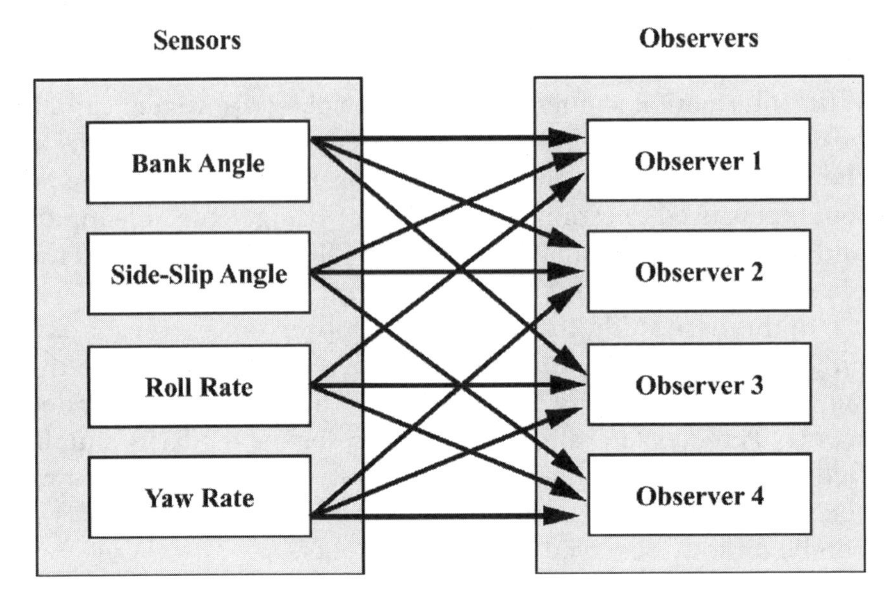

FIGURE 13-4 The observer scheme

only the observer—which does not receive that sensor's outputs—can estimate states properly. Therefore, by constructing the observer scheme in this manner, we can detect and identify a sensor's fault.

Each observer generates a residual by using the estimation errors as follows. The residual is a signal such that the feature of the fault is emphasized.

$$r_i = \sqrt{\frac{e_{i1}^2 + e_{i2}^2 + e_{i3}^2}{3}} \tag{13.1}$$

r_i: The residual of observer i ($i = 1 \sim 4$)

e_{ij}: The j-th normalized estimation error of observer i ($i = 1 \sim 4, j = 1 \sim 3$)

Residuals are usually small. When a sensor fault occurs, three residuals that use the fault sensor output become large. Only one residual that does not use the fault sensor output stays small. Thus, we can identify the fault sensor by monitoring four residuals.

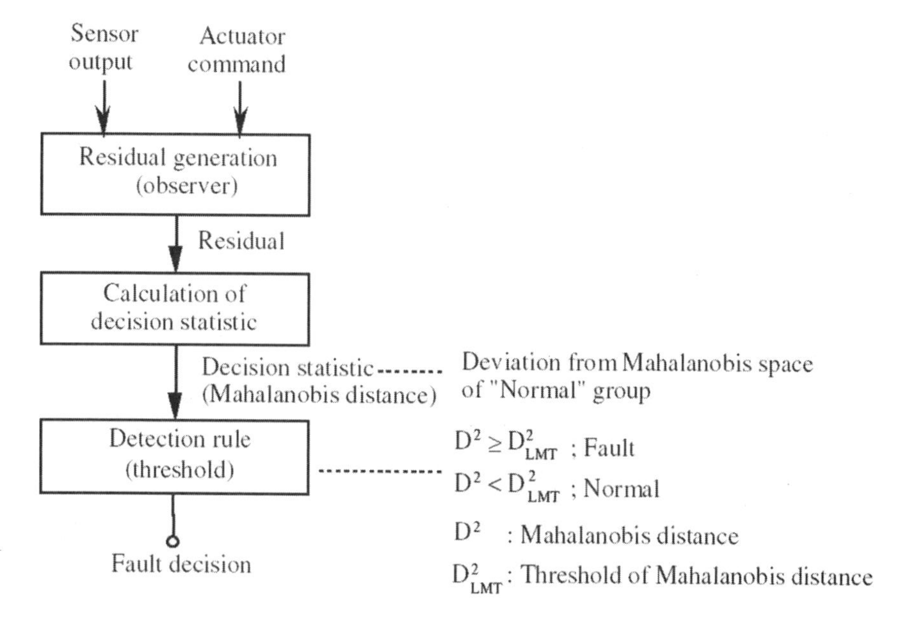

FIGURE 13-5 Process of fault diagnosis

13.4 PROCESS OF FAULT DIAGNOSIS BY USING RESIDUALS

Clearly, some correlation exists in the fluctuation of sequential residual data due to a sensor fault, although its statistical character is unknown. For this reason, MTS was adopted for manipulating sequential residual data.

The "normal" condition was defined. The samples of residual data under the normal condition were collected, and a database (i.e., Mahalanobis Space) was constructed. The possibility to decide whether a new sample would be normal or abnormal was examined by calculating its Mahalanobis Distance.

The process of fault diagnosis is shown in Figure 13-5.

13.5 COLLECTION OF NORMAL DATA

We added the aerodynamics errors to the dynamics model, and we changed the bank angle command. Then, we simulated normal

No.	$\delta C_{l\delta a}$	$dC_{l\delta r}$	$\delta C_{n\delta a}$	$\delta C_{n\delta r}$	σ_C
1	1	1	1	1	1
2	2	2	2	2	2
36	3	2	3	1	2

$\delta C_{l\delta_a}$, $\delta C_{l\delta_r}$, $\delta C_{n\delta_a}$, $\delta C_{n\delta_r}$: Aerodynamics error

σ_C : Bank angle command

Levels of aerodynamics error

	1 st	2 nd	3 rd
level 1	$-\sigma$	-2σ	-3σ
level 2	0	0	0
level 3	$+\sigma$	$+2\sigma$	$+3\sigma$

Levels of bank angle command

level 1	1.0 deg
level 2	1.5 deg
level 3	2.0 deg

FIGURE 13-6 Level definitions of "normal" group and assignment to L_{36}

cases and collected residual data. The aerodynamics error within 3 sigma and the bank angle commands from 1 degree to 2 degrees were defined as normal. Five factors—four aerodynamics errors and a bank angle command—were assigned to an L_{36} orthogonal array from the left column in order. Because the number of samples is expected to be larger than the number of data dimensions, we used L_{36} three times and obtained 108 samples of normal data. The level definitions each time and the assignment to L_{36} are shown in Figure 13-6. Bank angle commands were assumed to be a constant, or step input, for simplification.

13.6 THE STANDARD SPACE OF A NORMAL GROUP

We performed six seconds of simulation in order to obtain residuals. For each observer (1 ~ 4), by picking up 20 points samples at 0.3-second intervals from six seconds of residual data, we generated a vector of 80 dimensions. Mahalanobis Space was calculated by collecting normal data of 80 dimensions (refer to Figure 13-7).

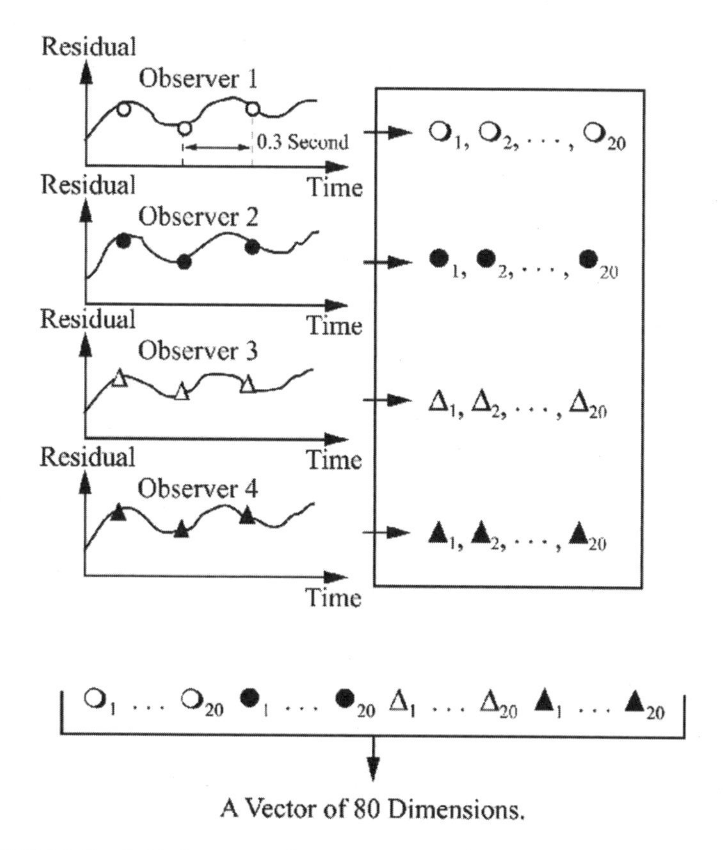

FIGURE 13-7 A 80-dimensional vector generated by six seconds of simulation

13.7 FAULT DETECTION BY MAHALANOBIS DISTANCE

The fault cases were simulated by multiplying the sensor output, such as zero or two. The Mahalanobis Distances then were calculated. The results are shown in Figure 13-8. For reference, Mahalanobis Distances of aerodynamics error cases (i.e., sensors that do not have faults but aerodynamics models that have −4 or −5 sigma errors) are also shown in the figure. This approach could detect some faults that other methods could not detect. Figure 13-8 shows Mahalanobis Distances of sensor

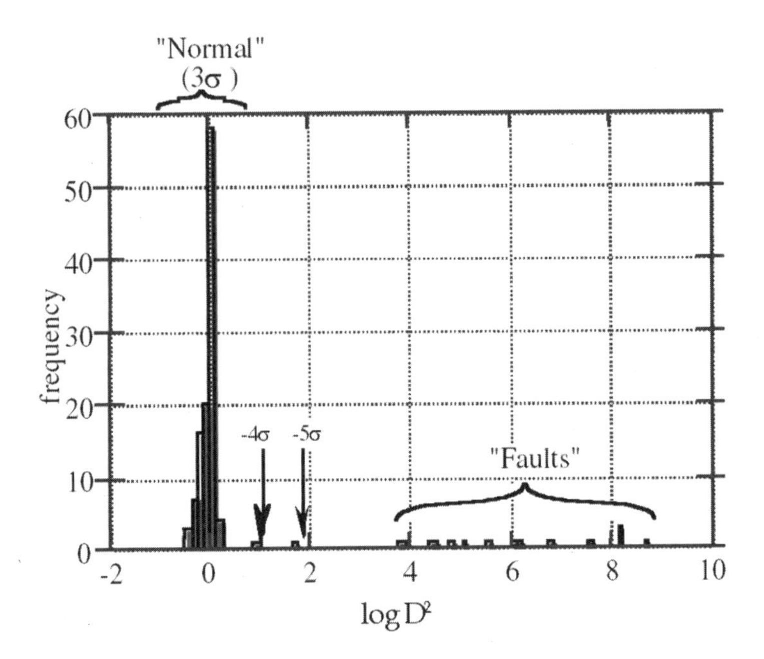

FIGURE 13-8 Frequency distribution of Mahalanobis Distance

fault cases that are large enough to distinguish them from those in the normal group.

13.8 FAULT IDENTIFICATION BY MAHALANOBIS DISTANCE

Next, we examined the possibility of fault identification by calculating Mahalanobis Distances for each observer. The expected fault-identification scheme by Mahalanobis Distance is shown in Table 13-1.

The results are shown in Table 13-2. The Mahalanobis Distance of Observer 1 could not be calculated, because the inverse of the correlation matrix could not be calculated. Also, in the small deviated fault cases (i.e., Nos. 8 ~ 13 of β sensor fault cases), differences in Mahalanobis Distances of Observers 2 ~ 4 did not appear clearly.

Therefore, you should detect a fault by using Mahalanobis Distance, which is calculated by all observers at first. Also, if a

TABLE 13-1 Fault-Identification Scheme by Mahalanobis Distance

MD content of fault	Observer 1	Observer 2	Observer 3	Observer 4
σ	large	large	large	small
β	large	large	small	large
p	large	small	large	large
r	small	large	large	large
normal	small	small	small	small

TABLE 13-2 Fault Identification by Mahalanobis Distance

No.	σ Simulation case	Observer 2	Observer 3	Observer 4
1	σ sensor fault (multiplied by 0)	14338.5	162776.0	35.5
2	// (multiplied by 2)	6392.8	148228.4	71.8
3	β sensor fault (multiplied by 0)	71.8	13.3	252.7
4	// (multiplied by 2)	3478.4	526.1	117476.6
5	p sensor fault (multiplied by 0)	52.2	122413.3	159625.6
6	// (multiplied by 2)	3.8	2494.4	722.3
7	r sensor fault (multiplied by 0)	37315.0	49006.6	129228.7
8	β sensor fault (multiplied by 0.9)	0.6	0.8	6.3
9	// (multiplied by 1.1)	1.4	0.9	23.4
10	// (multiplied by 0.8)	1.2	1.1	49.8
11	// (multiplied by 1.2)	4.9	2.8	93.4
12	// (multiplied by 0.5)	5.9	2.7	200.2
13	// (multiplied by 1.5)	82.7	44.0	1453.0
14	Aerodynamics error −4σ	1.3	1.4	3.2
15	Aerodynamics error −5σ	2.1	2.2	4.4
16	normal (out of use for making the standard space)	0.4	0.4	0.7
17	normal (in use for making the standard space)	0.5	0.4	0.4

fault is detected, then the fault sensor should be identified by using Mahalanobis Distances for each observer.

The fault of No. 8 in Table 13-2 could not be identified. We suspected that some information regarding the fault was lost when the estimation error was converted into a residual. Also, the inverse of the correlation matrix of Observer 1 could not be calculated, probably due to high correlative items in the correlation matrix.

Because the selection of characteristics for the standard space might cause these aforementioned problems, we decided not to select residuals; rather, we selected the estimation errors as items and tried the fault diagnosis again.

13.9 FAULT IDENTIFICATION BY ESTIMATION ERRORS

Some fault cases could not be well identified by using residuals. We tried three methods in order to use the estimation errors as follows. In all methods, 10-point data at 0.03-second intervals were considered as one group, and the mean was calculated to represent the 0.03-second interval. For earlier identification, the data within 30 points after the fault occurrence (i.e., about one second) were used (refer to Figure 13-9).

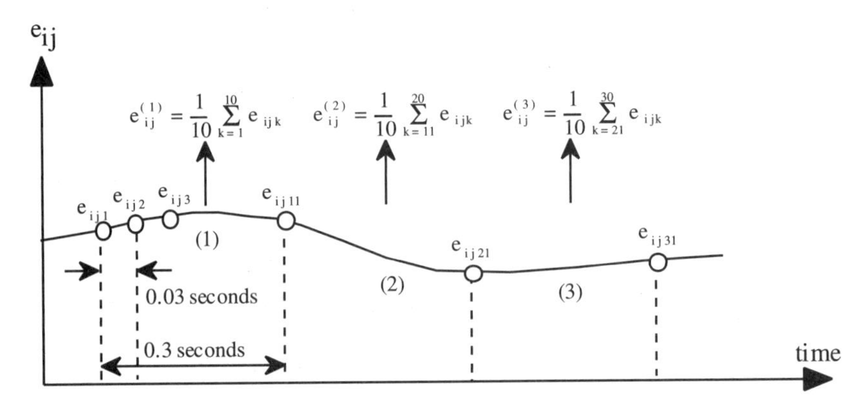

FIGURE 13-9 Methods of using estimation errors

By using any method of (A), (B), or (C), all fault cases in Table 13-2 could be identified along with the problems that resulted when residuals that were used did not occur. The results of these calculations are shown in Tables 13-3 through 13-5. "No." in the tables is the same as that in Table 13-2.

Next, we evaluated the functionality of three standard spaces made by methods (A) through (C). Because sensor fault cases (except the β sensor) had been clearly identified, we paid attention to only the β sensor faults. Because the distances of Observers 1, 2, and 4 are expected to become large in proportion as the fault value increases, we selected the difference between the β sensor fault value and the normal value as the signal factor.

On the other hand, the distance of Observer 3 is expected to stay small if the β sensor fault occurs, so we treated this characteristic as the smaller-the-better characteristic. The SN ratios of Methods (A) through (C) were compared by the total of Observers 1 through 4. The results are shown in Table 13-6. The signal factor and the square root of Mahalanobis Distances

TABLE 13-3 Fault Identification (Method A)

No.	Observer 1	Observer 2	Observer 3	Observer 4
1	97737.4	68970.5	24553.6	0.6
2	95933.9	61872.4	20660.1	0.5
3	3542.4	5730.7	0.5	10228.8
4	3713.9	9778.3	0.6	13866.9
5	110384.2	4.3	43821.1	59041.7
6	65649.8	0.9	12088.0	23833.3
7	19.9	115741.5	1573.4	65507.6
8	41.1	72.9	0.5	125.2
9	32.4	80.2	0.5	114.7
10	153.4	285.6	0.5	477.7
11	137.6	324.3	0.5	479.5
12	912.7	1651.3	0.5	2792.6
13	900.8	2170.8	0.5	3195.0
14	3.8	3.2	3.4	2.4
15	5.8	4.7	5.1	3.5
16	0.3	0.3	0.3	0.3
17	0.2	0.1	0.1	0.2

TABLE 13-4 Fault Identification (Method B)

No.	Observer 1	Observer 2	Observer 3	Observer 4
1	5706.4	565.3	1028.0	0.7
2	5496.4	499.7	934.9	0.5
3	53.4	223.2	0.5	845.7
4	56.9	168.1	0.4	650.0
5	5162.7	0.5	642.2	2230.1
6	3482.6	0.5	466.5	1312.7
7	2.8	1627.1	77.8	6139.6
8	0.5	4.1	0.4	12.1
9	1.5	0.8	0.4	4.1
10	1.6	11.6	0.4	38.8
11	3.6	4.9	0.4	22.2
12	12.0	58.8	0.4	214.1
13	16.3	39.7	0.4	160.3
14	0.7	0.6	0.5	0.9
15	0.9	0.7	0.6	1.1
16	0.5	0.5	0.5	0.6
17	0.0	0.0	0.0	0.0

TABLE 13-5 Fault Identification (Method C)

NO.	Observer 1	Observer 2	Observer 3	Observer 4
1	71455.6	137582.9	42839.3	1.2
2	69827.1	116030.7	28440.8	1.1
3	5294.3	7088.4	0.6	17803.3
4	5971.7	16208.5	0.7	33494.8
5	109784.4	7.7	135973.7	270939.1
6	48725.0	2.0	20430.1	81191.2
7	26.8	225507.1	2813.5	232713.9
8	63.3	104.7	0.5	258.3
9	51.1	126.0	0.5	255.2
10	237.0	411.3	0.5	980.4
11	217.5	508.9	0.5	1071.2
12	1397.3	2281.5	0.5	5451.2
13	1433.8	3481.2	0.6	7367.3
14	3.9	3.0	3.4	1.9
15	5.8	4.5	5.0	2.8
16	0.3	0.3	0.3	0.3
17	0.1	0.1	0.1	0.3

for Observers 1, 2, and 4 are shown in Figure 13-10. In Figure 13-10, the markers and the lines indicate the measured or computed data and the ideal functions, respectively. From Table 13-6 and Figure 13-10, we find that Method (A) is best.

TABLE 13-6 SN Ratio of Methods (A) through (C)

	SN Ratio (db)				
Method	observer 1	observer 2	observer 3	observer 4	Total
A	45.7	33.1	2.9	39.4	121.1
B	26.8	26.3	3.6	29.5	86.3
C	46.4	29.3	2.8	32.4	110.9

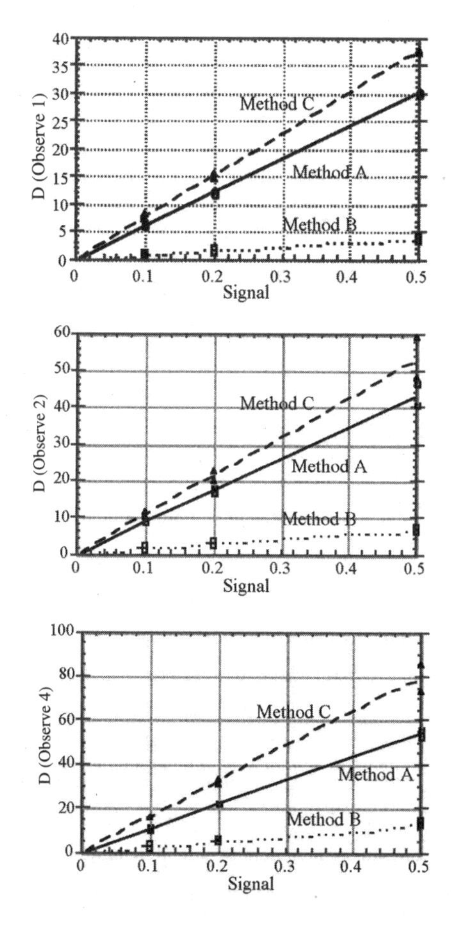

FIGURE 13-10 Signal factor and square root of Mahalanobis Distances

SECTION SIX

SOFTWARE INDUSTRY

CHAPTER FOURTEEN

VALUATION OF A PROGRAMMER'S CAPABILITY

14.1 INTRODUCTION

Finding a commodity that is not integrated with software is difficult. Software is used within televisions, refrigerators, and telephones and is used at banks, libraries, and government offices. Software is even used to purchase a train ticket, a telephone card, and more. Although software is used in various areas, no standard evaluation system for the software exists. One reason is because people produce software. Logically speaking, if we can assess people's abilities, we should be able to evaluate the software. The evaluation attempt is not only restricted to just software, however. Great effort has been made to quantify and distinguish different people's abilities for a long time. This task is not easy, however. Taguchi introduced the effectiveness of the *Mahalanobis Taguchi System* (MTS) by evaluating people's capabilities. In this chapter, we describe the attempt to distinguish people's programming capabilities from the survey data by using the MTS method.[1]

[1]Takada, Kei (Seiko Epson Corp.), Kazuhito Takahashi (University of Electro-Communication), Hiroshi Yano (Ohken Associates). "The Study of Valuation of Programmer's Capability by MTS Method," Quality Engineering Forum Symposium, 1998.

14.2 DATA COLLECTION

Researchers asked each subject to complete a survey and to write a simple computer program. In this study, 83 subjects participated. The sample consisted of a wide range of people—computer programmers, people who had little programming knowledge, college students, and technical school students. The survey consisted of 56 questions that were somewhat related to computer programming. Researchers asked the subjects to select one of the seven ratings for each question: strongly disagree (−3), disagree (−2), tend to disagree (−1), neutral (0), tend to agree (1), agree (2), and strongly agree (3). For programming, the researchers asked the subjects to write a simple computer program (for example, Base + Ball = Game, where one number is substituted for each letter and the same number is substituted for the same letter to complete the algebra). Researchers collected the data in order to assess the relationship between the pattern of the answer in the questions and how people will write a computer program.

14.3 THE MTS METHOD

In the Mahalanobis Space for this study, the distance of a person who belongs to the reference space is close to 1, and the distance of a person who does not belong to the reference space is large. The MTS method does not apply Mahalanobis Distance as used in discriminant analysis. Rather, the MTS technique judges the distance from the reference space (Mahalanobis Space) and evaluates the effectiveness of the factors by using the orthogonal array and the SN ratio.

In this study, we chose the reference space as "people without programming capability" (refer to Figure 14-1). The reason for this selection is because of the sample size. A small number of people in the world have programming capabilities. Thus, the sample size is limited. Based on this experience, people who have programming capabilities also have some independent characteristics. Therefore, it is preferable to select people who have programming capability as a reference group if the sample size is large.

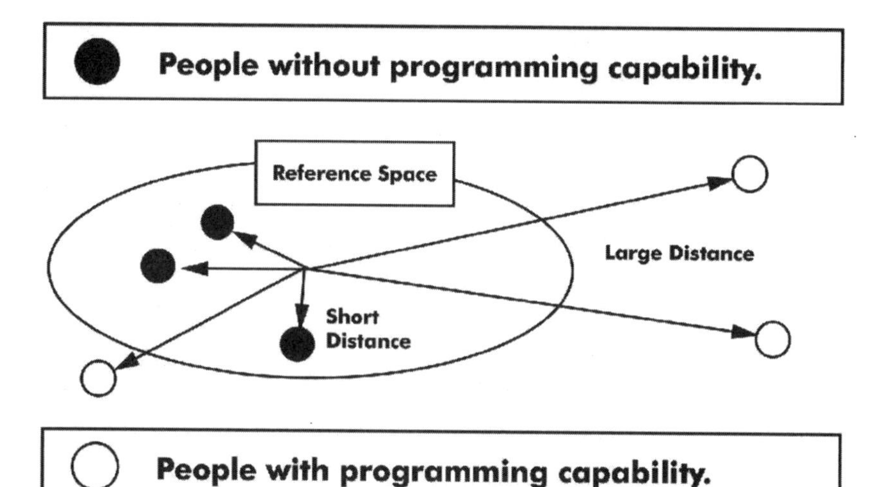

FIGURE 14-1 Reference space

The calculation of Mahalanobis Distance is as follows. Each data point is represented as Y_{il}, where i is the ith question (i, j = 1,2, . . . , k) and l represents the lth data (l = 1,2, . . ., n). First, calculate the average (m_i) and standard deviation σ_i from Y_{il}. Then, standardize Y_{il} by using the following equation:

$$y_{il} = \frac{Y_{il} - m_i}{\sigma_i} \text{ (i, j = 1,2,...k} \quad l = \text{1,2,...n)} \tag{14.1}$$

Next, calculate a correlation matrix R.

$$R = \begin{bmatrix} 1 & r_{12} & \dots & r_{1k} \\ r_{21} & 1 & \dots & r_{2k} \\ \dots & \dots & \dots & \dots \\ r_{k1} & r_{k2} & \dots & 1 \end{bmatrix} \tag{14.2}$$

where

$$r_{ij} = r_{ji} = \frac{1}{n} \sum_{l=1}^{n} y_{il} y_{jl} \text{ (i,j = 1,2,...k)} \tag{14.3}$$

Calculate the inverse matrix A from the matrix R. Matrix A is the reference space.

$$A = \begin{bmatrix} a_{11} & a_{12} & \cdots & a_{1k} \\ a_{21} & a_{22} & \cdots & a_{2k} \\ \cdots & \cdots & \cdots & \cdots \\ a_{k1} & a_{k2} & \cdots & a_{kk} \end{bmatrix} = \begin{bmatrix} 1 & r_{12} & \cdots & r_{1k} \\ r_{21} & 1 & \cdots & r_{2k} \\ \cdots & \cdots & \cdots & \cdots \\ r_{k1} & r_{k2} & \cdots & 1 \end{bmatrix}^{-1} \quad (14.4)$$

Using the elements from the reference space, you can calculate the Mahalanobis Distance as follows:

$$D^2 = \frac{1}{k} \sum_{i,j=1}^{k} a_{ij} y_i y_j \quad (i,j = 1,2,\ldots k) \quad (14.5)$$

We applied this calculation to the data that we obtained from the survey. In the analysis, we used the numerical values that we obtained from the questions ranging from -3 to 3. We also gathered some other characteristics from the survey, such as age, gender, years of programming experience, and years of personal computer experience. As far as gender, we categorized men as 1, and we categorized women as 2. Including the characteristics mentioned here, we used a total of 60 questions in the survey.

14.4 ANALYSIS

In the study, if a program was free of bugs and was easy to understand, then we categorized the person who wrote the program as "people with programming capability." We categorized 17 people out of 83 as "people with programming capability," and the other 66 people were categorized as "people without programming capability." Using the 66 subjects in the category, we formed a reference space.

The histogram of Mahalanobis Distance is shown in Figure 14-2. If we select an appropriate cutoff point, we can differentiate between the two groups. Thus, it is reasonable to use Mahalanobis Distance to analyze the survey data.

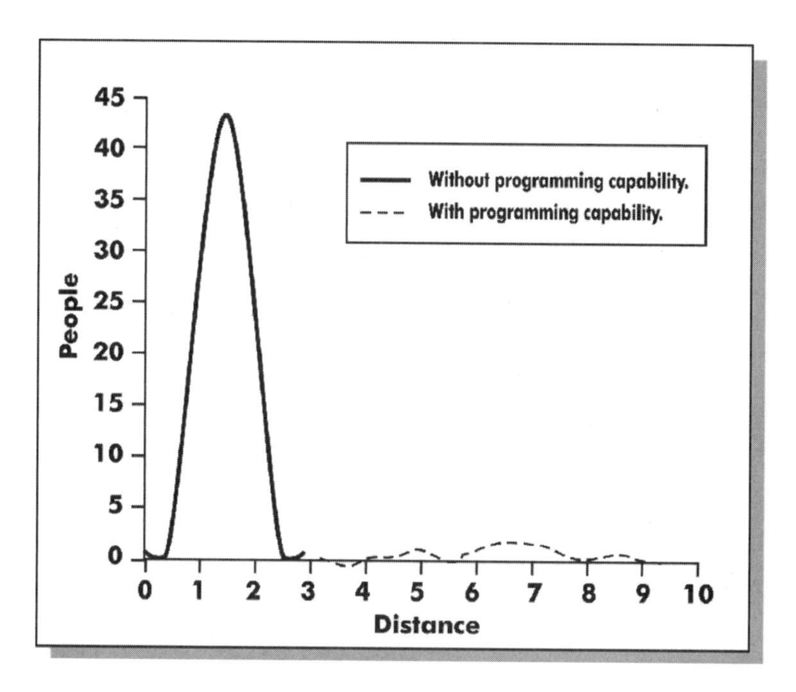

FIGURE 14-2 Distance from reference space (without programming space)

A need exists, however, to verify the power of discrimination when evaluating people's programming capabilities. In other words, we must assess the reliability of the constructed reference space.

When data is abundant—for example, we used 285 data elements out of 300 normal data elements to form a reference space—we calculated Mahalanobis Distances from the remaining normal data. If the distance is distributed around 1, then the reference space is reliable. If the distance becomes larger, this value indicates that either the number of data is insufficient or that problems exist with the selected questions.

When data is scarce (such as 50, for example), even removing five data points from a reference space will reduce the study's reliability. Therefore, we removed only a single data point for the study of reliability. We created a reference space by using 49 data points, and we calculated the distance from a single point. Next, we removed a different data point from the reference space and put back the previous data point in order to recalculate the reference space. This process determined whether the removed single data point was well reflected by the other 49 data points. From this point, we require some judgment experience. When this data is approximately 50 percent overlapping with the reference space, the data is reliable. If the data is 100 percent overlapping with other data points, the other data points represent the data point as well. If so, we can reduce the sample size. The result is shown in Figure 14-4. Because the sample size is small, both groups ("people with programming capability" and "people without programming capability") are separate from the reference space. Thus, this reference space cannot be used for evaluation.

14.5 FACTOR REDUCTION

The reason for the large distance that we calculated from the data that represented "people without programming capability" is because there were some unnecessary questions present in the survey. The task of finding unnecessary characteristics and removing them from the reference space is referred to as factor reduction. For factor reduction, you should treat each question as a factor and assign it to a Level 2 orthogonal array. Because there are 60 characteristics in a survey, we use an L_{64} orthogonal array.

Level 1 represents the situation in which a characteristic is used to form a reference space. Level 2 is a situation in which we do not use the characteristics. Calculate the distance from the abnormal data—the "people with programming capability." From the abnormal data, it is better to have a large distance. Thus, we use the larger-the-better SN ratio.

$$\eta = -10 \cdot \log \frac{1}{n} \left(\sum_{i=1}^{n} \frac{1}{D_i^2} \right) \quad (14.6)$$

Using the response graphs, select the most effective factors (questions) in distinguishing the people who have programming capabilities from the people who do not have programming capabilities. Level 1 represents data in the reference space, and Level 2 represents data that is not in the reference space. Thus, the line joining Level 1 and Level 2 from the upper left to the lower right (\ ) is an effective factor. The other factor with the trend (/) is harmful to the evaluation. The result of the calculation and the main effects are shown in Figures 14-3a through 14-3d.

From Figures 14-3a through 14-3d, all factors show the correct trend (\ ), indicating that the factors are effective in evalu-

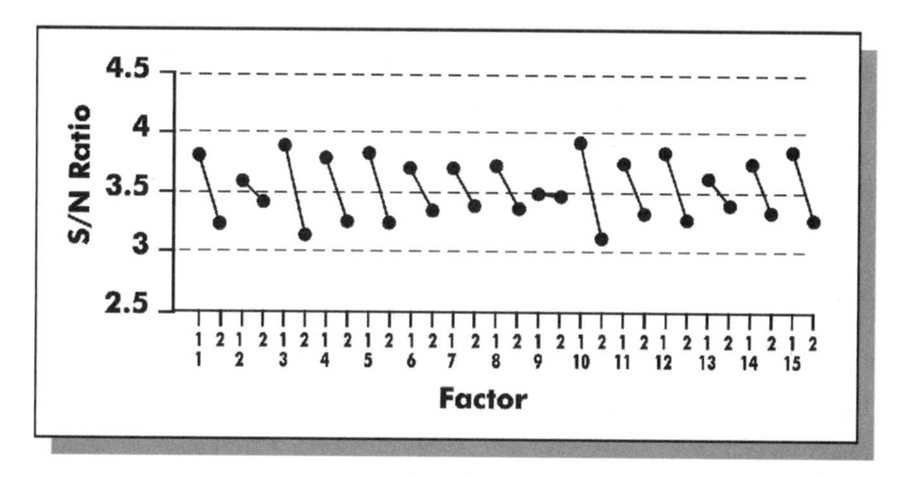

FIGURE 14-3a Factor reduction main-effect plot (Questions 1–15)

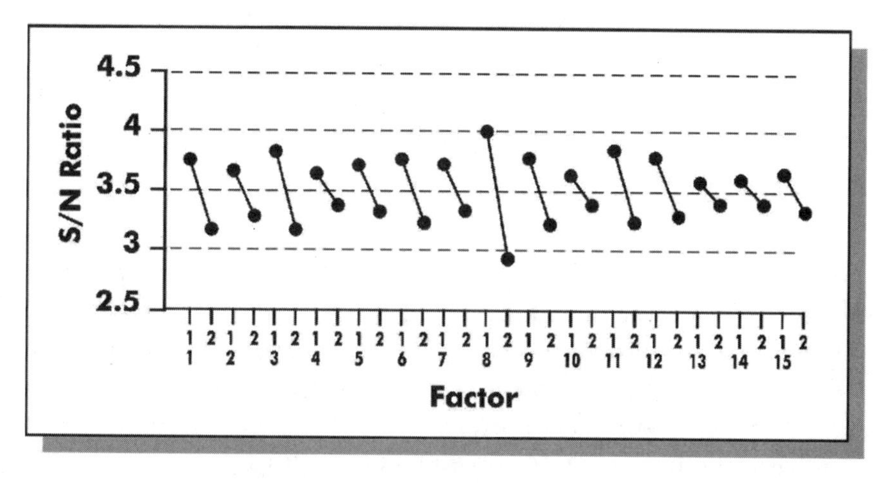

FIGURE 14-3b Factor reduction main-effect plot (Questions 16–30)

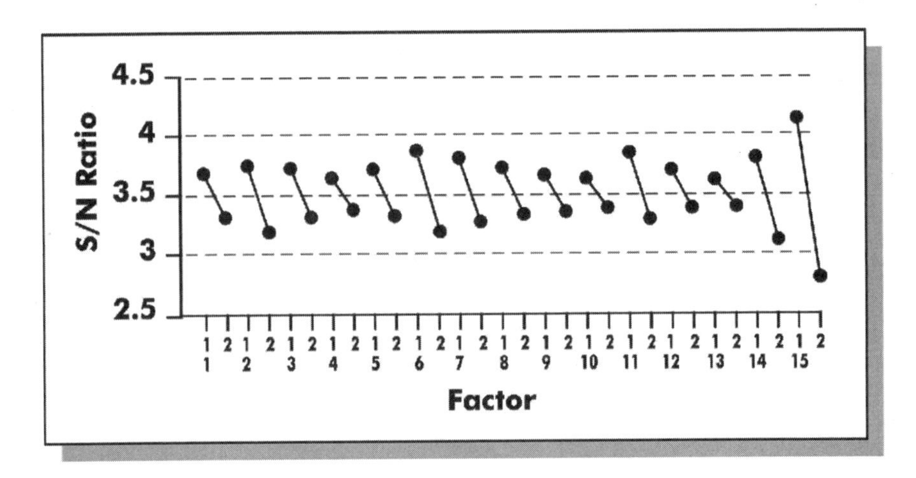

FIGURE 14-3c Factor reduction main-effect plot (Questions 31–45)

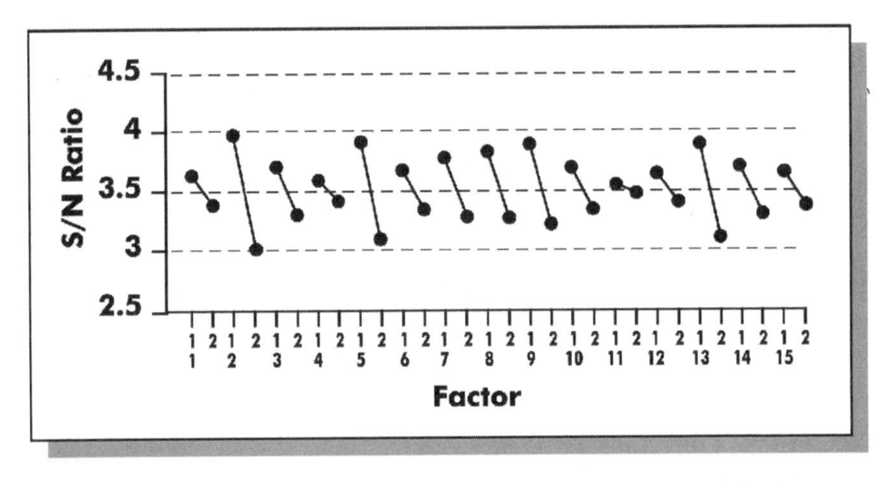

FIGURE 14-3d Factor reduction main-effect plot (Questions 46–60)

ation. We removed the factor with the smallest effect (Factor 9). Then, we performed factor reduction again. Question 13 showed the opposite trend (/) this time. We removed this factor, and we enlisted Factor 9 again. Factor 9 has the correct trend in the main-effect plot, although the effect is small. (After applying the factor reduction step seven times, Factor 9 showed the opposite trend and was removed.)

After the third round of factor reduction, there was a factor with the opposite trend (/) each time. Each time, we removed the factor from the reference space and repeated this process until 50 factors remained in the space. The remaining 50 factors showed the correct trend (/). Then, we fixed the first 20 factors, and remaining factors were assigned to an L_{32} orthogonal array. Using the different array, we selected the effective factors again. The number of factors are reduced from the reference space. Therefore, the distance from the abnormal data to the reference space also shortens. We must determine how close the distance from the abnormal data will become when a factor is reduced from the reference space. When the distance becomes too short, the factor might need to be brought back into the reference space.

Using this method, we reduced the 60 factors to 30. The result is shown in Figure 14-4. Data points for "people without

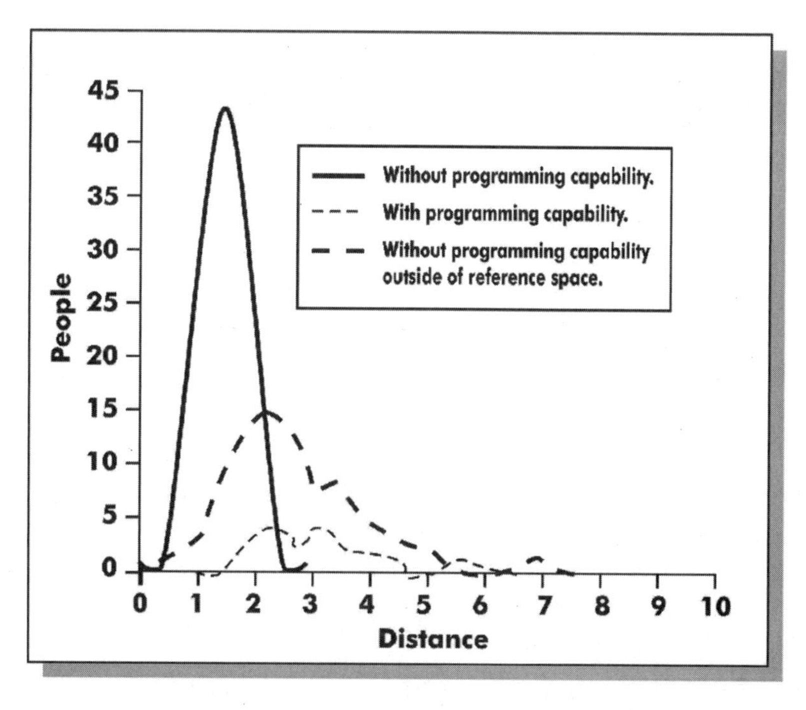

FIGURE 14-4 Reliability of data in the reference space after factor reduction

programming capability" who were not in the reference space overlapped with the reference space about 25 percent. To improve the percentage, the sample size has to be increased. After factor reduction, we can still differentiate between the two groups (the people who have programming capabilities and the people who do not have programming capabilities).

14.6 EFFECTIVE FACTORS

The effective factors are shaded in gray (refer to Table 14-1).

Next, the distribution of the data for people who have programming capabilities and people who do not have programming capabilities is summarized in Table 14-2. Based on each factor, we observed no major distinction between the two groups. Only

TABLE 14-1 Data for Figure 14-4

Distance	Capability		Without Capability Outside of Reference Space
	Without	With	
0	0	0	0
0.5	1	0	1
1	31	0	3
1.5	34	0	13
2	0	4	14
2.5	0	2	11
3	0	4	7
3.5	0	2	7
4	0	2	4
4.5	0	1	3
5	0	0	2
5.5	0	0	0
6	0	1	1
6.5	0	1	1
7>	0	0	0
Total	66	17	66

TABLE 14-2 Effective Factors

Question	Without Capability		With Capability		Average Difference
	Average	Std. Deviation	Average	Std. Deviation	
1	0.17	1.61	0.18	1.51	0.01
2	0.67	1.42	0.29	1.40	0.37
3	-0.05	1.55	0.12	1.76	0.16
4	2.12	0.92	1.76	1.09	0.36
5	0.41	1.53	-0.35	1.11	0.76
10	0.58	1.56	0.65	1.90	0.07
12	1.29	1.41	1.53	1.37	0.24
14	-0.47	1.60	-1.24	1.39	0.77
15	-0.08	1.69	-0.18	1.78	0.10
16	0.32	1.63	1.35	0.86	1.03
17	0.32	1.97	-0.06	2.14	0.38
19	-0.24	1.48	-0.18	1.55	0.07
20	0.53	1.25	0.29	1.53	0.24
21	0.77	1.41	0.40	1.22	0.37
22	1.44	1.45	1.14	1.50	0.30
23	0.85	1.42	0.29	1.36	0.55
24	0.41	1.93	0.53	1.81	0.12
26	0.39	1.50	0.76	1.52	0.37
27	-0.14	1.82	0.41	1.87	0.55
30	-0.08	1.71	-0.41	1.97	0.34
31	0.31	1.76	0.29	1.76	0.02
35	-1.09	1.57	-0.76	1.68	0.33
37	1.33	1.40	1.18	1.29	0.16
38	-0.03	1.69	-0.41	1.28	0.38
39	0.65	1.45	0.29	1.45	0.36
41	0.33	1.27	0.35	1.54	0.02
47	.75	1.19	0.24	1.52	0.52
50	0.67	1.14	0.29	1.05	0.37
51	-0.63	1.77	0.53	1.87	1.15
54	0.19	1.87	-0.53	1.94	0.72
55	0.19	1.58	0.18	1.38	0.01
56	-0.16	1.60	-0.06	1.78	0.10
58	1.12	0.33	1.00	0.00	0.12

when multiple characteristics are present can we distinguish one group from another.

14.7 RESULTS

We used the MTS method to evaluate the subjects' programming capabilities in this study, but this method can be used in other applications. The benefit of the MTS method is that once a reference space is established, people's capabilities can be evaluated only based on survey data. We can evaluate the capability even for people who currently have no knowledge in programming. Table 14-3 lists the survey questions.

TABLE 14-3 Survey Questions

Q1	If you are expecting an important announcement a week from now, does it affect your work?
Q2	Do you care about your appearance?
Q3	Do you clean up around you, or clean up PC files?
Q4	When you are doing something you love, do you lose track of time?
Q5	If your job function/content changes drastically from day to day, do you enjoy the change?
Q6	Does it bother you if your workplace is not neat?
Q7	If a project requires a full concentration for a week, do you like to take on the project?
Q8	Do you stick to the conventional method even when a new method is introduced?
Q9	When you start singing at Karaoke, do you monopolize the microphone?
Q10	Do you like puzzles? (jigsaw, Rubics cube, trick rings)
Q11	When the computer program works, do you add extra codes, or clean up the codes?
Q12	Do you enjoy working on a computer? (PC, work station)
Q13	Can you type without looking at the keyboard?
Q14	Do you like to talk in front of people?
Q15	Are you superstitious?
Q16	Do you like programming no matter what the project or content is?
Q17	Do you play TV games often?
Q18	Do you read comics often?
Q19	Do you follow fads?
Q20	When someone asks a question, do you like to take the time to explain?
Q21	Do you believe in something?
Q22	Do you like arts and crafts?
Q23	When something bothers you, do you get affected by it for a long time?
Q24	Do you enjoy sports programs on TV?
Q25	Are you late for an appointment often?
Q26	Do you like to read technical books?

TABLE 14-3 Survey Questions (*Continued*)

Q27	Can you stand simple tasks?
Q28	Do you like gambling? (such as Pachinko)
Q29	Do you enjoy cutting people down?
Q30	Do you like to draw or paint?
Q31	Do noisy surroundings bother you when you are in a train of thought?
Q32	Do you enjoy watching educational TV?
Q33	Do you enjoy working with people?
Q34	Do you imagine something out of the ordinary?
Q35	Do you forget to button your shirt?
Q36	Do you voice your opinion over others?
Q37	Do you like to know how a machine operates?
Q38	Does it bother you if someone makes a mistake on your name?
Q39	Are you a perfectionist?
Q40	Do you like science or logic?
Q41	Do you act on the spare of the moment?
Q42	Do you like to write? (novel, story)
Q43	Do you like to finish your job in a timely manner?
Q44	Does it bother you if someone is not dressed properly?
Q45	If you fail, do you continue to be bothered by it?
Q46	Do you believe you have programming capability?
Q47	Do you feel you are different from others?
Q48	Do you care about other peoples languages?
Q49	Are you confident about your health?
Q50	When you are on a job, does the immediate issue catch your attention?
Q51	Can you wake up early?
Q52	Are you good at calculating in your head?
Q53	Do you like to play sports?
Q54	Do you tend to get lost in a unfamiliar place?
Q55	Do you misspell or mistype words often?
Q56	Do you hesitate to go to a party?
Q57	Age?
Q58	Gender?
Q59	# of years in programming experience?
Q60	# of years using personal computer?

Chapter Fifteen

HANDWRITING RECOGNITION

15.1 INTRODUCTION

A human brain reportedly conducts pattern recognition from the information obtained from millions of optical eye cells, which are connected by a network of optical nerve cells. A high-level recognition, such as character recognition, requires learning and experience. Enabling a personal computer to conduct pattern recognition by using a system that is similar to the human brain is impossible, because the number of optical cells is so huge compared to the memory of computers—and the function of a brain is not available. In this study, we applied the concept of treating multi-dimensional information to hand-written character recognition.[1]

In order to construct a base space, we collected a number of hand-written characters of different styles, created by many different people. In this study, hiragana was used for recognition. Hiragana is one of the two sets of Japanese phonetic alphabets, having 50 characters (symbols) with cursive forms. One of the 50 hiragana characters was written in a 50-by-50 mm cell. We collected about 3,000 of these characters. Figure 15-1 shows three characters, pronounced "ah," "oh," and "nu," respectively.

We stored character "ah" as image data. The area in which a character was written was divided into n × n cells. The cells

[1]Kamoshita, Takashi, Ken-ichi Okumura, Kazuhito Takahashi, Masao Masumura, Hiroshi Yano. "Character Recognition Using Mahalanobis Distance," Quality Engineering Forum Symposium, 1997.

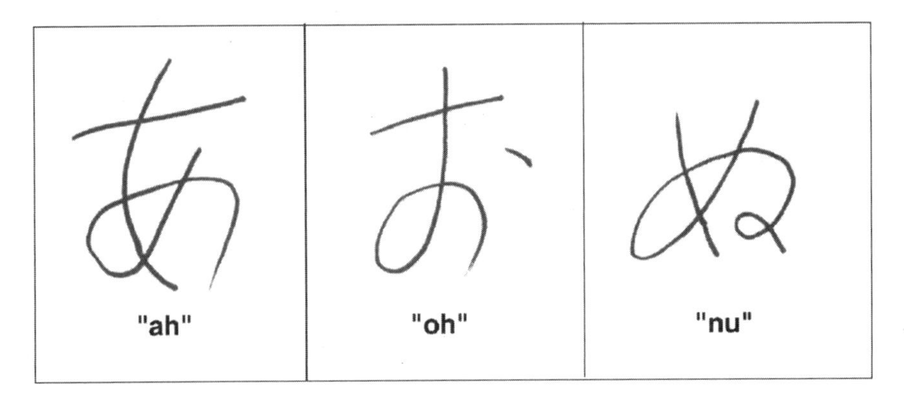

FIGURE 15-1 Handwriting of characters "ah," "oh," and "nu"

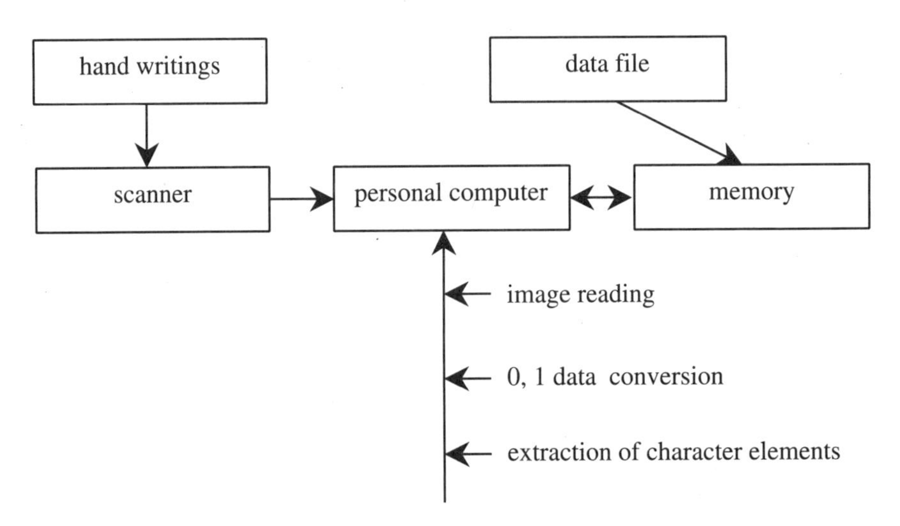

FIGURE 15-2 Preparation of characteristic elements

were digitized by assigning a 1 to the cells with images and 0 to the cells without images. When n is equal to 50, the number of characteristics is 2,500. This number is too large for personal computers to construct the inverse matrix. To reduce the number of characteristics, we used differential and integral methods (refer to Figure 15-2).

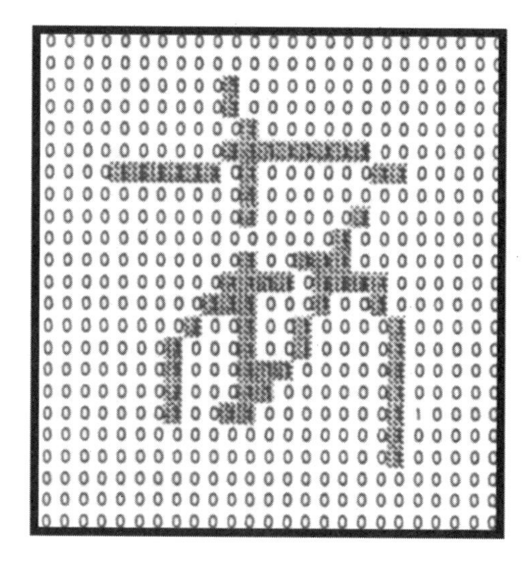

FIGURE 15-3 Digitization of the character "ah"

01 Data																										Integral (I)	Integral (II)	Differential
	0	0	0	0	0	0	0	0	0	0	0	0	0	0	0	0	0	0	0	0	0	0	0	0	0	0	0	0
	0	0	0	0	0	0	0	0	0	0	0	0	0	0	0	0	0	0	0	0	0	0	0	0	0	0	0	0
	0	0	0	0	0	0	0	0	0	0	1	0	0	0	0	0	0	0	0	0	0	0	0	0	0	1	1	1
	0	0	0	0	0	0	0	0	0	0	1	0	0	0	0	0	0	0	0	0	0	0	0	0	0	1	1	1
	0	0	0	0	0	0	0	0	0	0	0	1	0	0	0	0	0	0	0	0	0	0	0	0	0	1	1	1
	0	0	0	0	0	0	0	0	0	0	1	1	1	1	1	1	1	1	0	0	0	0	0	0	0	8	8	1
	0	0	0	0	1	1	1	1	1	1	0	1	0	0	0	0	0	0	1	1	0	0	0	0	0	9	16	3
	0	0	0	0	0	0	0	0	0	0	0	1	0	0	0	0	0	0	0	0	0	0	0	0	0	1	1	1
	0	0	0	0	0	0	0	0	0	0	0	1	0	0	0	0	0	1	0	0	0	0	0	0	0	2	7	2
	0	0	0	0	0	0	0	0	0	0	0	0	0	0	0	0	1	0	0	0	0	0	0	0	0	1	1	1
	0	0	0	0	0	0	0	0	0	0	0	1	0	0	1	1	1	0	0	0	0	0	0	0	0	4	6	2
	0	0	0	0	0	0	0	0	0	0	1	1	1	1	0	1	1	1	1	0	0	0	0	0	0	8	9	2
	0	0	0	0	0	0	0	0	0	1	1	1	0	0	0	1	0	0	1	0	0	0	0	0	0	5	10	3
	0	0	0	0	0	0	0	0	1	0	0	1	0	0	1	0	0	0	0	1	0	0	0	0	0	4	12	4
	0	0	0	0	0	0	0	1	0	0	0	1	0	0	1	0	0	0	0	1	0	0	0	0	0	4	13	4
	0	0	0	0	0	0	0	1	0	0	0	1	1	1	0	0	0	0	0	1	0	0	0	0	0	5	13	3
	0	0	0	0	0	0	0	1	0	0	0	1	1	0	0	0	0	0	0	1	0	0	0	0	0	4	13	3
	0	0	0	0	0	0	0	1	0	0	1	1	0	0	0	0	0	0	0	1	1	0	0	0	0	5	14	3
	0	0	0	0	0	0	0	1	1	1	0	1	0	0	0	0	0	0	0	0	1	0	0	0	0	5	14	3
	0	0	0	0	0	0	0	0	0	0	0	1	0	0	0	0	0	0	0	0	1	0	0	0	0	2	10	2
	0	0	0	0	0	0	0	0	0	0	0	0	0	0	0	0	0	0	0	1	0	0	0	0	0	1	1	1
	0	0	0	0	0	0	0	0	0	0	0	0	0	0	0	0	0	0	0	1	0	0	0	0	0	1	1	1
	0	0	0	0	0	0	0	0	0	0	0	0	0	0	0	0	0	0	0	0	0	0	0	0	0	0	0	0
	0	0	0	0	0	0	0	0	0	0	0	0	0	0	0	0	0	0	0	0	0	0	0	0	0	0	0	0
	0	0	0	0	0	0	0	0	0	0	0	0	0	0	0	0	0	0	0	0	0	0	0	0	0	0	0	0
Integral (I)	0	0	0	0	1	1	1	6	3	3	6	15	4	3	4	4	4	3	3	8	3	0	0	0	0			
Integral (II)	0	0	0	0	1	1	1	13	13	13	16	16	12	11	10	8	7	7	7	16	3	0	0	0	0			
Differential	0	0	0	0	1	1	1	2	3	3	4	2	3	3	3	2	2	3	2	3	1	0	0	0	0			

FIGURE 15-4 Elements extracted by using 25-by-25 cells

15.2 EXTRACTION OF CHARACTER ELEMENTS

We divided character "ah" into 25-by-25 cells, as shown in Figures 15-3 and 15-4. We extracted character elements by using differential and integral methods as follows:

(1) Integral Method (I): The number of cells with the image (or "1") from the start to the end in each row or in each column
(2) Integral Method (II): The number of cells occupied by the image (or "1") in each row or in each column
(3) Differential Method: The number of continued cells in each row or in each column. The values collected by using these methods are shown in Figure 15-4.

15.3 PROCEDURES OF CHARACTER RECOGNITION

The following list describes the procedure of character recognition:

(1) Collection of hand-written characters
(2) Scanning of characters
(3) Digitizing to 0 and 1
(4) Extraction of characteristic elements
(5) Construction of a base space
(6) Calculation of the Mahalanobis Distance
(7) Calculation of the Mahalanobis Distance of unknown characters
(8) Recognition of unknown characters

We collected hand-written characters from many people, without any modification. Those characters were scanned and saved as image data. Each image was divided into $n \times n$ cells and was transformed into 0 and 1 data bits in order to collect characteristic elements. From these data, we constructed a base space, and we calculated a Mahalanobis Distance from each hand-written character. Next, we collected a new group of hand-written characters that were not related to the previous base group. For recognition, we calculated the Mahalanobis Distances of these newly collected characters.

15.4 CALCULATION OF MAHALANOBIS DISTANCE

For the simplicity of explaining the calculation of Mahalanobis Distance, we decomposed character "ah" into 5-by-5 cells, as shown in Table 15-1. The image in the cells was digitized.

In the table, the shaded numbers show the results of using the integral method (I). These numbers consist of 10 elements, as shown in Table 15-2. For simplicity, we used only 20 characters of "ah." There are 10 elements and 20 hand-written characters.

Raw data is denoted by Y_{ij}. From each element, the average (m) and standard deviation (σ) are calculated to obtain normalized raw data, yij, where i denotes the element (i = 1,2, . . . ,10) and j denotes the number of characters (j= 1,2, . . . ,20).

For example, the normalized value of $y_{1.1}$ is calculated as follows:

$$y_{1.1} = \frac{Y_{1.1} - m_1}{\sigma_1} = \frac{4 - 3.3500}{1.19478} = 0.544033 \qquad (15.1)$$

TABLE 15-1 Digitized character image

Character "ah" No. 1							
						Integral (I)	Item Number
Digitized	0	0	1	0	1	3	No. 6
	1	1	1	1	1	5	No. 7
	0	1	1	1	1	4	No. 8
	1	1	1	0	1	5	No. 9
	1	1	1	1	1	5	No. 10
Integral (i)	4	4	5	4	5		
Item No.	1	2	3	4	5		

TABLE 15-2 Results of extracting 20 characters from "ah"

	Item (Characteristic)									
	Data					Data				
	Item No. 1	Item No. 2	Item No. 3	Item No. 4	Item No. 5	Item No. 6	Item No. 7	Item No. 8	Item No. 9	Item No. 10
No. 1	4	4	5	4	5	3	5	4	5	5
No. 2	5	5	5	5	5	5	3	4	5	5
No. 3	1	4	5	2	3	1	3	4	4	4
No. 4	5	5	5	5	3	4	3	4	5	5
No. 5	3	5	4	3	3	3	2	5	5	5
No. 6	5	5	5	4	5	5	1	4	5	5
No. 7	3	5	4	5	5	4	4	5	5	4
No. 8	3	5	5	5	4	3	5	5	5	4
No. 9	4	5	5	0	3	2	2	4	5	5
No. 10	2	5	5	5	3	3	3	5	4	4
No. 11	1	5	4	2	3	1	3	3	5	4
No. 12	3	4	4	4	3	1	4	4	5	3
No. 13	2	4	5	4	4	1	4	1	5	5
No. 14	4	5	5	3	3	3	4	4	5	5
No. 15	2	5	4	3	5	4	1	4	5	5
No. 16	4	4	5	5	5	5	4	5	5	3
No. 17	4	5	5	1	3	2	3	3	4	5
No. 18	4	5	5	4	3	2	4	4	4	5
No. 19	4	4	4	3	3	2	3	4	5	5
No. 20	4	5	4	2	3	1	4	4	5	5
Total	67	94	93	69	74	55	66	80	96	91
Average	3.3500	4.7000	4.6500	3.4500	3.7000	2.7500	3.3000	4.0000	4.8000	4.5500
Standard Deviation	1.19478	0.458258	0.47697	1.430909	0.9	1.373863	1.004988	0.894427	0.4	0.668954

Similarly, $y_{1.10}$ is calculated as follows:

$$y_{1.10} = \frac{Y_{1.10} - m}{\sigma_1} = \frac{2 - 3.3500}{1.19478} = 1.12991 \qquad (15.2)$$

The average and standard deviation of each element becomes 0 and 1, respectively. The normalized values are shown in Table 15-3.

Next, we calculate a correlation coefficient from each pair of elements. Then, we can construct their correlation matrix. Letting

$$s, t = 1, 2, \ldots, 10$$

and

$$j = 1, 2, \ldots, 20$$

the correlation coefficient, r, is calculated as follows:

$$r_{st} = r_{ts} = \frac{1}{20} \sum_{j=1}^{20} y_{sj} y_{tj} \qquad (15.3)$$

TABLE 15-3 Normalized results of characters “ah”

	Item (Characteristic)									
	Integral (I) (Vertical)					Integral (II) (Horizontal)				
	Item No. 1	Item No. 2	Item No. 3	Item No. 4	Item No. 5	Item No. 6	Item No. 7	Item No. 8	Item No. 9	Item No. 10
j=1	0544033	-1.52753	0.733799	0.384371	1.444444	0.181969	1.691563	0	0.5	0.672692
j=2	1.381007	0.654654	0.733799	1.083228	1.444444	1.637718	-0.29851	0	0.5	0.672692
j=3	-1.96689	-0.52753	0.733799	-1.01334	-0.77778	-0.27378	-0.29851	0	-2	-0.82218
j=4	1.381007	0.654654	0.733799	1.083228	-0.77778	0.909843	-0.29851	0	0.5	0.672692
j=5	-0.29294	0.654654	-0.36277	-0.31449	-0.77778	0.181969	-0.29355	1.118034	0.5	0.672692
j=6	1.381007	0.654654	0.733799	0.384371	1.444444	1.637718	-2.28859	0	0.5	0.672692
j=7	-0.29294	0.654654	-1.36277	1.083228	1.444444	0.909843	0.696526	1.118034	0.5	-0.82218
j=8	-0.29294	0.654654	0.733799	1.083228	0.333333	0.181969	1.691563	1.118034	0.5	-0.82218
j=9	0.544033	0.654654	0.733799	-2.41106	-0.77778	-0.54591	-1.2355	0	.05	0.672692
j=10	-1.12991	0.654654	0.733799	1.083228	-0.77778	0.181969	-0.29851	1.118034	-2	0.82218
j=11	-1.96689	0.654654	-1.36277	-1.01334	-0.77778	-1.27378	-0.29851	-1.11803	0.5	-0.82218
j=12	-0.29294	-1.52753	-1.36277	0.384371	-0.77778	-1.27378	0.696526	0	0.5	-2.31705
j=13	-1.12991	-1.52753	0.733799	0.384371	0.333333	-1.27378	0.696526	-3.3541	0.5	0.672692
j=14	0.544033	0.654654	0.733799	-0.31449	-0.77778	0.181969	0.696526	0	0.5	0.672692
j=15	-1.12991	0.654654	-1.36277	-0.31449	1.444444	0.909843	-1.29355	0	0.5	0.672692
j=16	0.544033	-1.52753	0.733799	1.083228	1.444444	1.637718	0.696526	1.118034	0.5	-2.31705
j=17	0.544033	0.654654	0.733799	-1.7122	-0.77778	-0.54591	-0.29851	-1.11803	-2	0.672692
j=18	0.544033	0.654654	0.733799	0.384371	-0.77778	-0.54591	0.696526	0	-2	0.672692
j=19	0.544033	-1.52753	-1.36277	-0.31449	-0.77778	-0.54591	-0.29851	0	0.5	0.672692
j=20	0.544033	0.654654	-1.36277	-1.01334	-0.77778	-1.27378	0.696526	0	0.5	0.672692
Total	-1E -14	-2.3E-14	1.55E-14	1.55E-15	-3.1E-14	1.33E-15	-8E-15	-9.8E-15	0	1.22E-14
Average	-4E-16	-1.2E-15	7.77E-16	7.77E-17	-1.5E-15	6.66E-17	-4E-16	-4.9E-16	0	6.11E-16
Standard Deviation	1.025978	1.025978	1.025978	1.025978	1.025978	1.025978	1.025978	1.025978	1.025978	1.025978

For example, $r_{1.1}$ is

$$r_{1.1} = \frac{1}{20}\left[0.54033^2 + \ldots\ldots + 0.54033^2\right] = 1 \tag{15.4}$$

The correlation matrix, denoted by R, is as follows:

$$R = \begin{bmatrix} r_{1.1} & r_{1.2} \cdots\cdots, r_{1.10} \\ r_{2.1} & r_{2.2} \cdots\cdots, r_{2.10} \\ \cdots & \cdots\cdots\cdots\cdots \\ r_{10.1} & r_{10.2} \cdots\cdots, r_{10.10} \end{bmatrix}$$

$$= \begin{bmatrix} 1.00000 & 0.19177 & 0.30270 & \ldots & 0.38473 \\ 0.19177 & 1.00000 & -0.02288 & \ldots & 0.37514 \\ \cdots & \cdots & \cdots & \cdots & \cdots \\ 0.38743 & 0.37514 & 0.1332 & \ldots & 1.00000 \end{bmatrix} \tag{15.5}$$

From the correlation matrix, we can construct its inverse matrix (denoted by A).

$$A = \begin{bmatrix} a_{1.1} & a_{1.2} \cdots\cdots, a_{1.10} \\ a_{2.1} & a_{2.2} \cdots\cdots, a_{2.10} \\ \cdots\cdots & \cdots\cdots \\ a_{10.1} & a_{10.2} \cdots\cdots, a_{10.10} \end{bmatrix}$$

$$= \begin{bmatrix} 2.4485 & 0.8424 & -0.2777 & \ldots & -0.9766 \\ 0.4824 & 1.8126 & 0.3819 & \ldots & -0.794 \\ \cdots & \cdots & \cdots & \cdots & \cdots \\ -0.9766 & -0.7640 & 0.0571 & \ldots & 2.15720 \end{bmatrix} \quad (15.6)$$

The Mahalanobis Distance is given by the following equation:

$$D_j^2 = \frac{1}{10} \sum_{st=1}^{10} a_{st}\, y_{sj}\, y_{tj} \quad (15.7)$$

When j = 1, its Mahalanobis distance is calculated as follows:

$$D_1{}^2 = \frac{1}{20}[2.4485 \times 0.544033^2 + \ldots\ldots + 2.1572 \times 0.672692^2]$$

$$= 1.041664064$$

15.5 HAND-WRITTEN CHARACTER RECOGNITION

In this study, we only discuss the recognition of the character "ah." Fifteen-hundred characters of "ah" were collected for the construction of the base space. A separate group of 100 characters of "ah" were prepared. Also, 100 each of characters "oh" and "nu," which look like character "ah," were prepared. The Mahalanobis Distances of those characters were calculated from the base space, as shown in Figure 15-5.

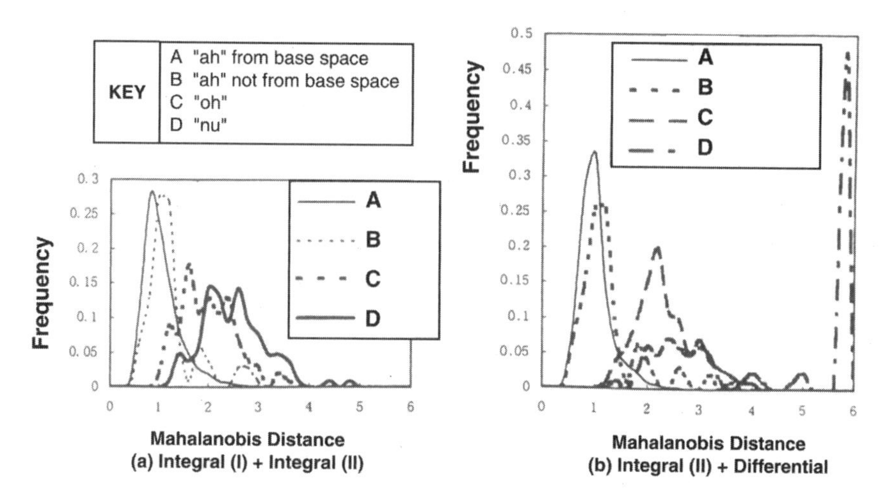

FIGURE 15-5 Distributions of different characters

In the figure, the abscissa shows the Mahalanobis Distance, and the ordinate shows frequency. The Mahalanobis Distances have an average that is equal to 1 and distribute symmetrically. The newly collected character "ah" is almost overlapped in the base space. Although the characters "oh" and "nu" slightly overlapped with the base space, these characters are seemingly distinguishable. Figure 15-5 (b) shows the results of using integral (II) and differential methods, indicating a better discrimination.

Figure 15-6 shows the Mahalanobis Distance versus cumulative frequency. We can see that when the Mahalanobis Distance is less than 1, the character can be recognized as "ah." But there are about 20 percent of characters "ah" that are above a threshold of 1.5. Above threshold 3, there are almost no characters "ah."

For the characters "oh" and "nu," none of their Mahalanobis Distances are less than 1. Twenty percent of the characters "oh" have a Mahalanobis Distance of less than 2, and 10 percent of the character "nu" have the same distance. After this study, character recognition was conducted for those characters that do not look like "ah." No problems existed with recognizing these characters, however.

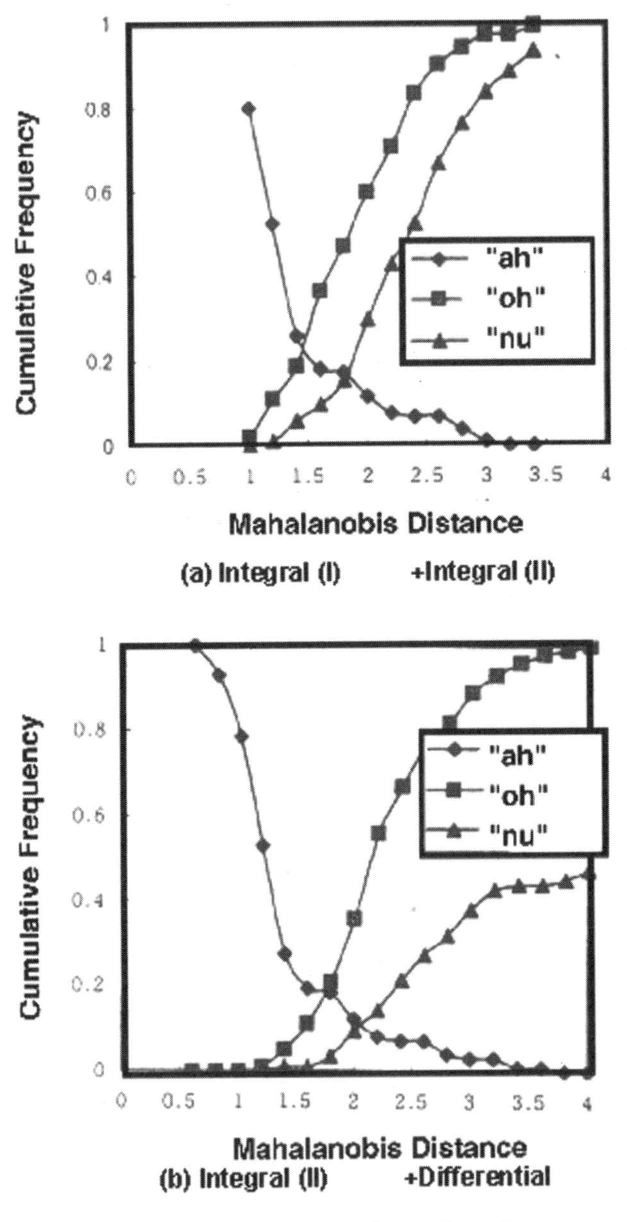

FIGURE 15-6 Mahalanobis Distance and cumulative frequency

Section Seven

GOVERNMENT

CHAPTER SIXTEEN

U.S. DOLLAR BILL INSPECTION

16.1 PATTERNS OF U.S. DOLLAR BILLS

The face values of U.S. dollars include $1, $2, $5, etc., with the portraits of George Washington, Alexander Hamilton, Andrew Jackson, etc. on the front. All bills have the same dimensions, and values are judged by recognizing the patterns of the portraits. Reportedly, the portraits have many repeating patterns. In the scanning process, sensors scan a portrait (as shown in Figure 16-1) to obtain the curves of concentration changes, as shown in Figure 16.2.[1]

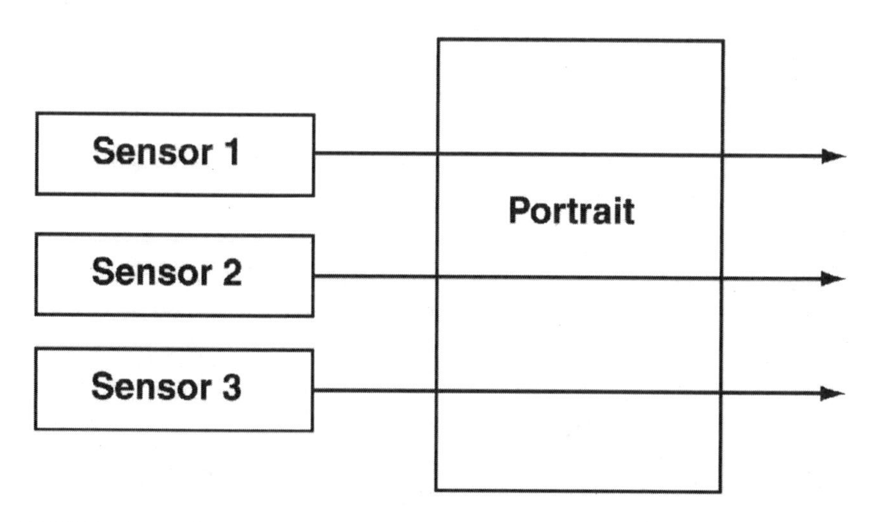

FIGURE 16-1 Scanning of a portrait

[1]Sakano, Susumu. "U.S. Dollar Bill Inspection," Quality Engineering Forum Symposium, 1998.

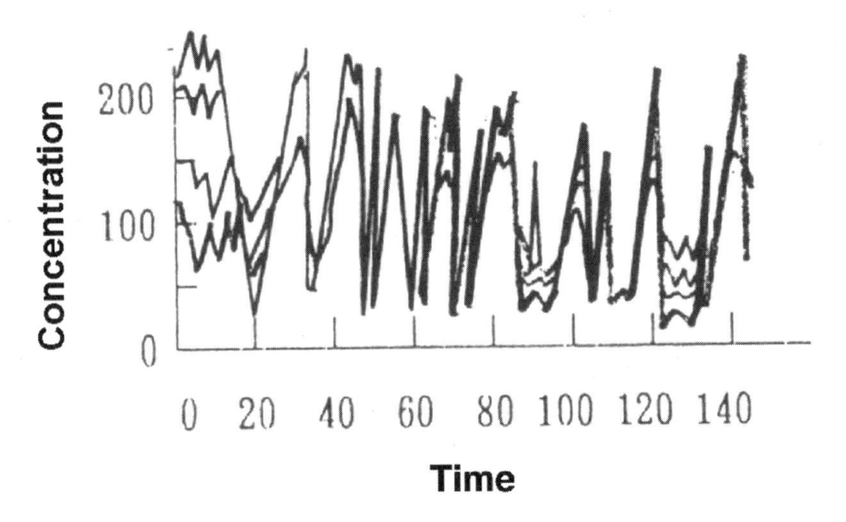

FIGURE 16-2 Image concentration

You can use different methods to quantify the curves after scanning:

(1) Use differential and integral characteristics (Figure 16-3)
(2) Use the Fourier coefficient (Figure 16-4)
(3) Use Fourier transformed spectral analysis (Figure 16-5)

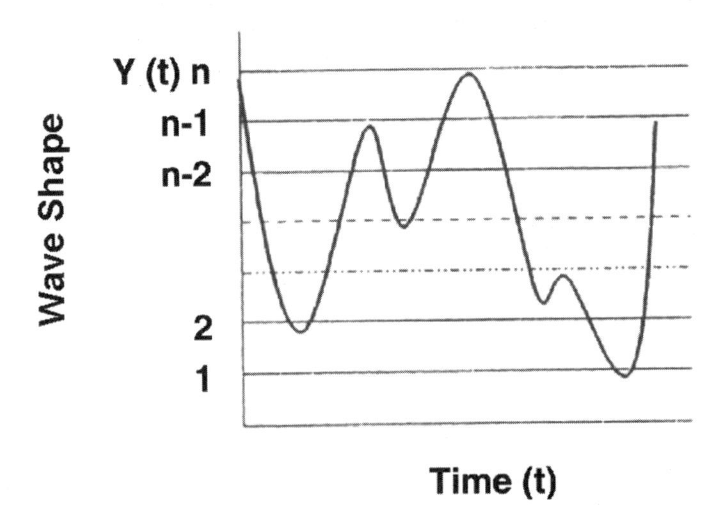

FIGURE 16-3 Differential and integral characteristics

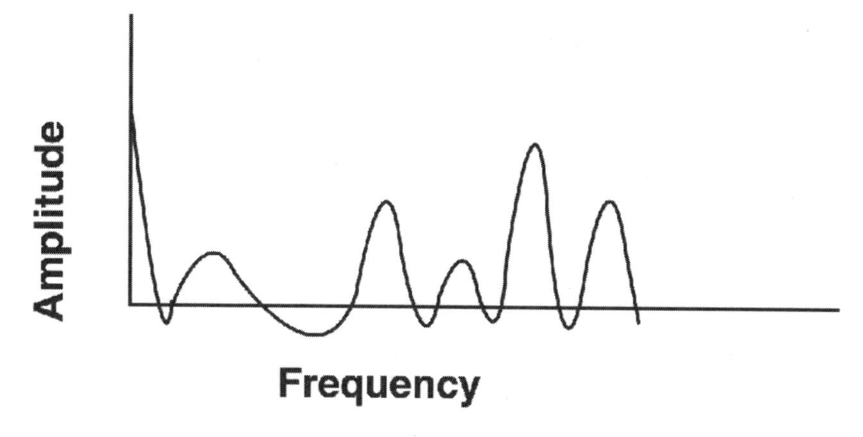

FIGURE 16-4 Fourier coefficient transformation

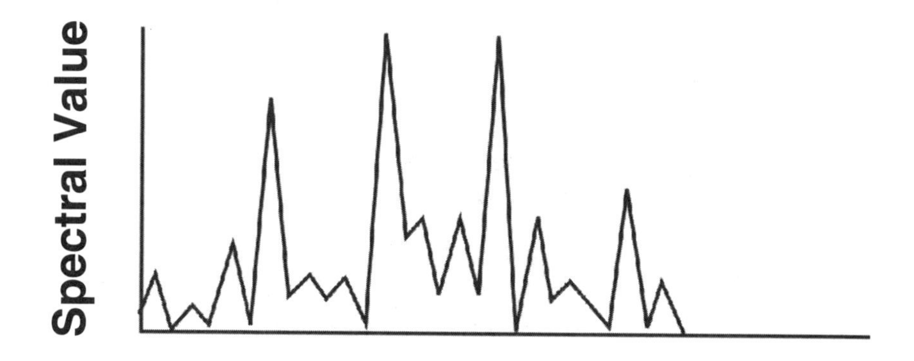

FIGURE 16-5 Fourier transformed spectral analysis

The first method—the differential characteristic—relates to the type of information with regard to frequency, integral characteristic, and amplitude. The second method is convenient for quantifying the periodic trend. The third method is known for observing the spectral distribution.

In portraits, photographs, paintings, or printed matter, one would expect the drastic gradation change to not be significant, but there are many repeating patterns. In this study, we used the Fourier coefficient transformation.

16.2 CHARACTERISTICS OF ONE-DOLLAR BILLS

The scanned time-series data was transformed to Fourier coefficients in order to obtain the amplitude/frequency curves, as shown in Figure 16-6.

From Figure 16-6, we see that there are nine projections in a one-dollar bill that characterize this type of bill. We expect that counterfeit bills might have different projections; e.g., the amplitude might be smaller. Therefore, the amplitudes of nine projections of 10 one-dollar bills were used to construct a Mahalanobis Space (refer to Table 16-1).

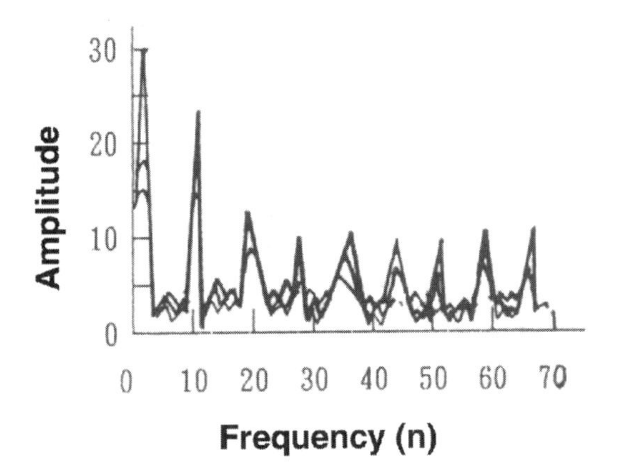

FIGURE 16-6 Amplitude of Fourier coefficient in a one-dollar bill

TABLE 16-1 Raw Data

	Y_1	Y_2	....	Y_9
Note No.1	$Y_{1.1}$	$Y_{2.1}$	....	$Y_{9.1}$
2	$Y_{1.2}$	$Y_{2.2}$	....	$Y_{9.2}$
....	....			
10	$Y_{1.10}$	$Y_{2.10}$	....	$Y_{9.10}$

After normalizing this data, we constructed a Mahalanobis Space for the calculation of Mahalanobis Distance.

16.3 DIFFERENTIATION OF ONE-DOLLAR BILLS

Because no counterfeits of one-dollar bills were available for the experiment, genuine one-dollar bills were modified and were used for the experiment. Researchers modified the bills by altering some areas in the portrait to make them blank or creating a folded line in order to distort the portrait. Figure 16-7 shows the results of differentiation. In the figure, solid dots represent genuine bills, and hollow circles represent modified bills. You can differentiate between other dollar bills by using the same approach.

In this study, we constructed the Mahalanobis Space by using the nine items in the Fourier coefficient transformed image concentration. Differentiation could be simplified, however, if the number of items could be reduced. For this purpose, we used an L_{12} orthogonal array to assign the items. In the array, we assigned Level 1 and Level 2 as "use" and "do not use," respectively. From the combination of "use" and "do not use" of each run, we calculated the larger-the-better SN ratio. Table 16-2 shows the layout and SN ratios, and Table 16-3 shows the response table.

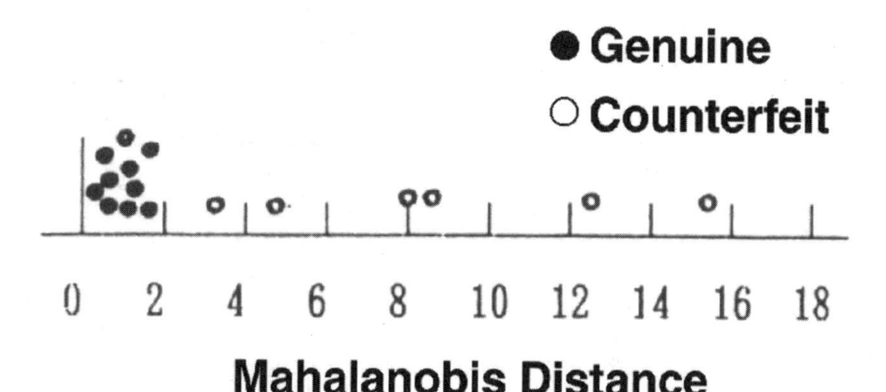

FIGURE 16-7 Differentiation of one-dollar bills

TABLE 16-2 Layout and SN Ratios

No. \ Item	1	2	3	4	5	6	7	8	9	S/N Ratio (db)
1	1	1	1	1	1	1	1	1	1	15.1176
2	1	1	1	1	2	2	2	2	2	16.8399
3	1	1	2	2	2	1	1	1	2	17.3672
4	1	2	1	2	2	1	2	2	1	17.0434
5	1	2	2	1	2	2	1	2	1	16.7796
6	1	2	2	2	1	2	2	1	2	18.3582
7	2	1	2	2	1	1	2	2	1	11.6617
8	2	1	2	1	2	2	2	1	1	11.8464
9	2	1	1	2	2	2	1	2	2	12.4911
10	2	2	2	1	1	1	1	2	2	3.4840
11	2	2	1	2	1	2	1	1	1	3.2606
12	2	2	1	1	2	1	2	1	2	2.2671

Table 16-3 shows that only items 1 and 2 contribute to the differentiation, while the others negatively contribute. In other words, the other seven items should not be used. From items 1 and 2, we constructed a Mahalanobis Space, and the results of differentiation are shown in Figure 16-8.

TABLE 16-3 Response Table

Item	Level 1	Level 2	Difference
1	16.9177	7.5018	9.4159
2	14.2207	10.1988	4.0219
3	11.1700	13.2495	-2.0795
4	11.0558	13.3637	-2.3079
5	11.4537	12.9658	-1.5121
6	11.1568	13.2626	-2.1058
7	11.4167	13.0028	-1.5861
8	11.3695	13.0500	-1.6805
9	12.6128	11.8013	0.8170

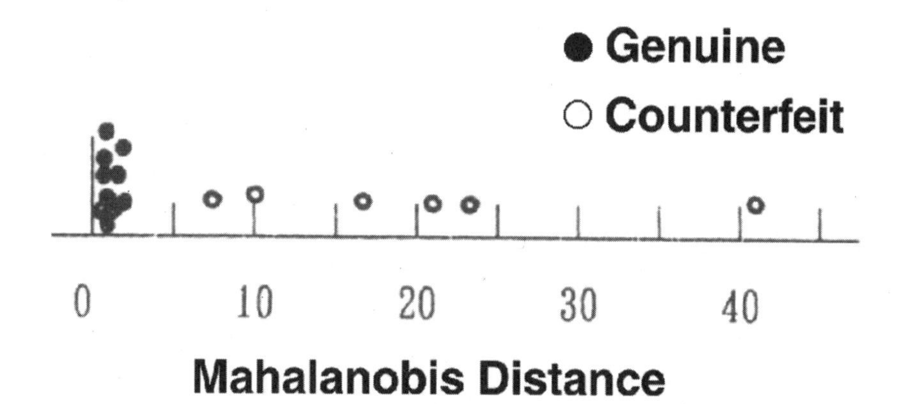

FIGURE 16-8 Discrimination of one-dollar bills by using two items

INDEX

D

E–F

N

O–P

Q

R

S

T

V–W

ABOUT THE AUTHORS

GENICHI TAGUCHI, D.Sc., is Executive Director of the American Supplier Institute, an international organization that trains and consults on quality management and engineering. Winner of the Deming Prize of Japan and Shewhart Medal of American Society for Quality, an inductee into the Automotive Hall of Fame, and winner of many other notable awards, Dr. Taguchi developed the statistical and logical systems for rapid improvements in product quality and product development that have revolutionized automobile and other industries in America and in his home country of Japan. He is the co-author of McGraw-Hill's *Robust Engineering* and an honorary member of ASQ and ASME.

SUBIR CHOWDHURY is Executive Vice President of the American Supplier Institute. He has been popularizing fire prevention methodology among senior management worldwide. He is a recipient of the Society of Automotive Engineers' most prestigious Henry Ford II Award for Excellence, and was also awarded by the U.S. Congress and the Automotive Hall of Fame. Mr. Chowdhury's major expertise is in the field of management and leadership. His most recent international best-selling business book *Management 21C: Someday We'll All Manage This Way* is the lead business title of Financial Times-Prentice Hall. This most valuable work is a major contribution in the field of management of the 21st century and is praised by best business thinkers and CEOs of the world. He is the co-author of *Robust Engineering* (McGraw-Hill, 1999).

YUIN WU is Executive Director of the American Supplier Institute and conducted the first Taguchi Methods® experiments in the United States while working in private industry in California. He also provides consultation and training on Robust Engineering and MTS in North America as well as many countries in Europe, South America, and Asia. In addition, Mr. Wu has trained for the automotive, computer and defense companies as well as mechanical, electrical, chemical and food industries. He has written numerous publications on the subject of Taguchi Methods®. He is co-author of *Taguchi Methods® for Robust Design®* (ASME Press, 2000).